GROUNDSWELL

ALSO BY EZRA LEVANT

Ethical Oil: The Case for Canada's Oil Sands
*Shakedown: How Our Government Is Undermining Democracy
in the Name of Human Rights*

Ezra Levant

GROUNDSWELL

The Case for Fracking

SIGNAL

McCLELLAND
& STEWART

Copyright © 2014 by Ezra Levant

Signal is an imprint of McClelland & Stewart, a division of Random House
of Canada Limited, a Penguin Random House Company.

All rights reserved. The use of any part of this publication reproduced,
transmitted in any form or by any means, electronic, mechanical,
photocopying, recording, or otherwise, or stored in a retrieval system, without
the prior written consent of the publisher – or, in case of photocopying or other
reprographic copying, a licence from the Canadian Copyright Licensing
Agency – is an infringement of the copyright law.

Library and Archives Canada Cataloguing in Publication

Levant, Ezra, 1972-
    Groundswell / Ezra Levant.

ISBN 978-0-7710-4644-5

    1. Hydraulic fracturing. 2. Gas well drilling. 3. Shale gas.
4. Shale oils. I. Title.

TN880.2.L49 2013   622'.3381   C2013-900697-4

Library of Congress Control number: 2013901599

Typeset in Electra

Printed and bound in the United States of America

McClelland & Stewart,
a division of Random House of Canada Limited
A Penguin Random House Company
One Toronto Street
Suite 300
Toronto, Ontario
M5C 2V6

www.randomhouse.ca

1   2   3   4   5   18   17   16   15   14

CONTENTS

GROUNDSWELL

I magine if the United States suddenly discovered oil, tens of billions of barrels of oil, maybe even enough oil to rival Saudi Arabia. It wouldn't just make America richer. It would change the world.

Producing that oil would mean hundreds of thousands of new jobs in the American oil industry, and more indirectly everywhere. And then imagine if that happened in Canada. And Poland. And Ukraine. And a dozen other democracies.

Adding that much oil production to the world's supply would also mean the price would fall, and so the price of everything made from oil would fall too, like gasoline and chemicals and plastic. And so would the cost of everything that runs on oil, not just what we use in our cars, but everything shipped on a truck or a plane.

But the real transformation would be political. America would no longer be importing nine million barrels of oil a day from foreign countries – forking over about a billion dollars a day, much of it to hostile regimes like Saudi Arabia or Venezuela. In fact, those same tanker terminals currently importing foreign oil might even be turned around, if America produced a surplus, to ship U.S. oil to the world. Europe's democracies, especially former Soviet bloc countries, would finally be truly free of Russia's domination.

So it wouldn't just be about the cash. It would be about ending the insecurity of importing oil from dictators. It would be about freedom.

Well, all that just happened in America. And it's about to happen in Europe, too.

Not with oil, but with its hydrocarbon cousin, natural gas, the cleanest-burning fossil fuel, the fuel used to heat home furnaces and cook on stovetops, and increasingly to fire up electrical power stations. All of a sudden, in the past ten years, America "discovered" it had staggering amounts of natural gas resources – 664 trillion cubic feet more – that just weren't there before. If you're wondering, that's like adding twenty-six[1] years of supply at the current U.S. rate of consumption (on top of conventional gas resources that would last sixty-five years).

But here's the miraculous part: the new gas was always known to be there. Until now, it just wasn't possible to recover it.

Conventional oil and gas production is like drilling into

the centre of a watermelon, putting in a straw, and sucking out the juice. It works only when oil or gas is pooled in a porous rock, like rainwater soaking into a sandbox. And once you pump out the good stuff, you're done.

But enormous quantities of oil and natural gas aren't in pools, ready to be pumped out. They're trapped, in tiny amounts, in the pores of fine-grained rocks called shale.

You can't drill for those droplets of gas or oil the conventional way. It would be like trying to use a straw to suck the juice from an apple.

But in 1947,[2] the Stanolind Oil and Gas Corporation – formerly J.D. Rockefeller's Standard Oil of Indiana – tried out an experiment to increase the production of a 2,400-foot-deep gas well in Grant County, Kansas. A thousand gallons of gelled gasoline were injected into the well under hydraulic pressure, to create little cracks in the limestone half a mile underground, with the hope of releasing the gas that was trapped in the porous rock. The process was called "Hydrafrac," and it was patented in 1949. The Halliburton Oil Well Cementing Company was given an exclusive licence to use the process.

In the first year, Halliburton fracked 332 wells, and the process increased the average production of the wells by 75 per cent. Over the next fifty years, the process was repeated approximately one million[3] times in the United States. But it wasn't until fracking was combined with another technology, horizontal drilling, that the fracking revolution took off – and that extra 664 trillion cubic feet of gas moved from fantasy to reality.

Horizontal drilling is just what it sounds like – to use the drinking straw analogy, it's a bendy straw, that goes straight down, but once it's deep underground it slowly curves until it's horizontal. That doesn't just let the well travel along a seam of oil or gas for a much longer distance, it also reduces the number of drilling rigs needed on the surface. Instead of having many wells, all spread out, going straight down into the same geological formation, now one small pad on the surface can have many horizontal wells drilling down and then radiating out in different directions, like daisy petals, covering an area that used to need many drilling rigs. Horizontal wells drilled on land can even reach out under the sea.

It was that technological combination – hydraulic fracturing plus horizontal drilling – along with other improvements such as high-tech mapping that led to the new fracking revolution. It started where you might guess it would: just west of Southfork, the legendary ranch that was the setting of *Dallas*, the hit TV series set in the oil and gas industry.

There was always oil and gas activity in that part of Texas, in the geological formation called the Barnett Shale, but even J.R. Ewing would have been stunned at the transformation of the industry. The first horizontal fracked wells in Barnett were drilled about ten years ago, timidly at first: 400 in 2004. But in 2010[4] alone, 10,000 fracked wells were drilled there, and now fracked wells account for 70 per cent of the production in the area.

The technology has been proven – in traditional oil and gas states like Texas, but now also in places that weren't traditionally thought of as big energy producers. The Bakken Shale formation, which straddles the border between North Dakota and Canada, has copied the exponential growth in Texas. Horizontal fracking there has made North Dakota the second[5]-largest oil-producing state in the United States, at 942,000[6] barrels a day. For comparison, that's more than three times as much oil as the OPEC country of Kuwait[7] now sells to America. It's also why North Dakota has the lowest unemployment rate in the United States, an impossibly low 2.7 per cent.

Fracking has revived America's oil industry. But it has positively revolutionized the natural gas industry, bringing so great a supply of natural gas onto the market that America is set to become a natural gas exporter – and the price of gas has plummeted to historic lows.

Fracking really has turned the United States into the Saudi Arabia of natural gas. The beautiful technology with the ugly name is the hottest job-creator in America and has reduced the cost of nearly everything that needs energy to run.

But not everyone is thrilled. Countries that currently dominate the natural gas industry – OPEC dictatorships like Iran and Qatar, and Vladimir Putin's Russia – don't like the new competition that's threatening to break their cartel and bring down prices for consumers. And the prospect of hundreds of years of clean, low-cost energy has become unhappy news for environmental extremists in the West, who were counting on high power prices to make green technologies like wind turbines and solar panels economically competitive.

This book looks at the promise of natural gas that fracking has made possible. But more importantly it looks at the enemies of fracking: who they are, what they're saying, and why they're fighting the future so desperately.

Chapter One

PLENTIFUL, CHEAP, CLEAN, DEMOCRATIC ENERGY

G eologists have long known that, spread out over vast areas of the globe, natural gas was trapped in tiny pockets in compressed mud, called shale. There was just no economic way to get that gas – until now. This is no miracle: it is a combination of market freedom and technology. With world energy demands rising every year, entrepreneurs had an equally rising incentive to exploit new energy resources. It was only a matter of time before some brilliant scientist figured out how to get all that trapped natural gas out of the earth a mile down, and make a fortune doing it. Sure enough, someone did.

The word "revolution" is tossed around casually, but it truly fits here: the technology has suddenly turned something that was economically useless and geologically meaningless into something valuable, as if sand itself had suddenly became valuable. And in an instant, great swathes of the Sahara Desert would have an economic future – presuming corrupt warlords didn't hijack it all.

For decades the world has been calling for some kind of miracle energy solution that would give us cheap, clean, widely available power. Nuclear power was part of that solution, but it couldn't be transported, it didn't prove to be as cheap as its visionaries had hoped, and environmentalists eventually came to oppose it with a passion equal to their loathing for fossil fuels.

Then we had the wild fantasy that was "cold fusion," which was another kind of nuclear reaction – so the environmentalists probably would have hated it, too. Cold fusion was supposed to make nuclear technology even more widely available and economical, but that turned out to be wishful thinking. More than twenty years after the scientific community thoroughly debunked the infamous Fleischmann-Pons experiment, which claimed to have produced cold fusion, it remains little more than a pipe dream.

Green activists have been demanding a new energy source and pressed governments in the United States and Europe to invest massive subsidies into wind and solar power. But those, too, have proven to be costly fantasies. The energy they produce is prohibitively expensive. And unlike nuclear power, they are inefficient and unreliable. Solar works, of course, only when there's sunshine, which rules out half the day and many months of the year in some regions. Wind works only when the wind blows, or doesn't blow too much. Both technologies require duplication: a back-up generating system, to fire up when they're not working. That just adds to the cost: instead of one power generator, you now need two. Some progress. Some solution.

But horizontal shale-gas fracking is progress. It is a solution. It's cheap. It's widely available. It's clean. But it didn't come out of some government-planned, spectacular moon-shot-type mega-project. There isn't much that's sexy about it. It wasn't created by famous geniuses, like Albert Einstein or Robert Oppenheimer. It was developed by a boring engineer making incremental, practical improvements on existing drilling technology, a private businessman almost no one has ever heard of, named George Mitchell.

In practice, fracking is a brief moment in the life of an oil or natural gas well. Once drilled, an oil or gas well can continue quietly pumping away for years, even decades. Fracking a well is usually done in a week's time.[1]

Fracking isn't done on all oil or gas wells; the geological nature of the rock has to call for it: compressed mud, called shale, usually a mile or so underground. A drilling rig is assembled – that takes a few days. And then a deep hole is drilled into the rock, and a steel casing is inserted and cemented into place. It's like putting a mile-long pipe into the ground. A separate, outside layer of steel pipe surrounds the well as it passes through any underground sources of drinking water, extending deep below it.

So far, none of that is fracking – it's just drilling. And the drilling continues deep through rock, often a mile or more, the equivalent of passing through a mountain. Once the hole comes close to the targeted layer of shale, the pipe is curved, like a bendy straw, to become horizontal. It can burrow sideways for up to two miles along a seam of hydrocarbons. It looks a bit like a foot – and in fact drillers call the front of the horizontal section the "toe" and the part that bends the "heel." That horizontal well is cemented into place too, all miles underground.

That's the well – but it's a closed steel pipe; nothing can go in or out of it. So a specialized "gun" is lowered down to perforate the pipe, creating little holes through the steel, cement, and nearby rocks.

The drilling is done; the drilling rig is removed. And if fracking hadn't been invented, that would be it – oil and gas would flow into the well and be brought to the surface. But that doesn't work if the oil and gas are trapped in pores of shale rock. So the fracking stage begins – usually taking three to ten days.

In early days, wild entrepreneurs tried to fracture rocky wells using explosives. Lt.-Col. Edward Roberts, a veteran of the U.S. Civil War, used his battlefield observations about artillery shells to develop a type of torpedo[2] that was lowered down the well to explode, cracking open the shale. These days, thankfully, high-pressure water is used instead. That's the "hydraulic" part of hydraulic fracturing.

Water is pumped into the well under high pressure (some new technologies use gas, like compressed air) to crack open the shale. Sand is mixed into the water, so that individual grains of sand prop open the little cracks in the rock. And a small number of chemicals are included, depending on the geology.

When the fracking is done, part of the fracking water is brought back up to the surface, to be recycled in other frack jobs or disposed of as per environmental regulations.

That's fracking – a brief stage in the life of a well. Once it's done, all the water trucks and equipment are moved away, and the surface site is reclaimed. All that's left is usually a nondescript little shack on top, with equipment to collect the gas that flows, sometimes for decades.

Fracking isn't the same as drilling for oil and gas. It's a technological process that enhances drilling for oil and gas. It's so effective that an estimated 90 per cent of all oil and gas wells in North America are fracked – in fact, old oil and gas fields, thought to have been depleted using pre-fracking technology, have often been given new life using fracking. The procedure has been used an estimated 1.1 million times in the United States alone, and hundreds of thousands of times in Canada, too.

Fracking is the difference between daydreaming about "miracle" science-fiction solutions and an industry-based approach to realistic, achievable solutions. Science fiction is called fiction for a reason: it's fun, but it's make believe. Playwrights have a Latin term for that: "*deus ex machina*" – literally "god from the machine," where God or an angel appears in the middle of a play (lowered onto the stage by a machine), to resolve an impossible problem with the plot. But life isn't a Christmas movie or Greek play. There is no magical solution to our energy challenge.

Fracking isn't some cheap plot gimmick. Natural gas isn't some rarefied, enchanted mineral with fantastical properties, like the precious "unobtanium" they were seeking in James Cameron's hit sci-fi film *Avatar*. It's more like the smartphone: an incredible breakthrough in technology that has helped improve our lives, but one that came about only after decades of laying down the foundation with the necessary preceding technologies – computers, microchips, the Internet, wireless technologies, and a thousand other step-by-step developments.

Shale-gas fracking isn't a *deus ex machina*, but the effect it is having comes awfully close to doing what that kind of plot contrivance does: it resolves so many of the energy challenges we've faced until now, it's difficult to overstate the implications.

The most incredible effect of fracking is the impact it has on the amount of energy we have in the world. With one technological leap, we have increased the world's gas reserves by 40 per cent, an astonishing magnitude. And it has opened up energy reserves in countries that have never had them before. Countries like Poland and Ukraine, which have had to depend on Russia, their hostile former occupier and autocratic petro-state, for their basic daily needs. Countries like Israel, surrounded by oil, all of it in the hands of enemies, but with virtually none of its own – a country that has lived the last six decades in a permanent state of existential insecurity, but also energy insecurity. The natural gas transformation is making energy-starved countries into energy powerhouses from scratch.

Before the shale-gas revolution arrived, fossil fuels were a bittersweet blessing: they helped our lives immensely, but they were largely controlled by the very regimes in the world that frequently made our lives worse. Russia, Saudi Arabia, Venezuela, Iran, Libya, Iraq – these are all places that have contributed to world instability (Russia and Iran still do), and the only reason they could have that kind of influence is that they were getting rich on their oil and gas racket.

Suddenly, though, it's not just OPEC and "gas OPEC" that have this vital energy that we all want and need. Now, so many friendly, liberal, democratic countries have rich and accessible deposits of fossil fuels, too. Fracking doesn't just change the balance of the world's energy clout, it can change the balance of the world's geopolitical power. It saps the strength of menacing regimes and empowers the good guys. It's a historic shift and one that's long overdue.

Just look at the effect on the most powerful "good guy" there is: the United States. The United States is a massive producer of oil and gas – the thirteenth-largest worldwide for oil and in the top three for gas production – but as much as it produces, its economy is just that much bigger. It's a massive energy importer. Shale gas comes with incredible economic potential for a country like the United States: it's not just the money that comes directly from the development of shale gas itself and the spinoff economic effects. Shale gas is also flooding America with cheap, easily available energy. Many American factories were relocated overseas because China or India could offer cheaper labour. But cheap energy can give America a new advantage. Before

2009, American natural gas prices were, at times, higher than either those of the U.K. or Japan.[3] For industries that require immense volumes of natural gas to power their energy-intensive processes, the United States was becoming an increasingly unaffordable place to do business. But as of 2009, because of the shale boom, U.S. prices started falling precipitously. Gas prices were falling everywhere – in the U.K. and Japan, too. But only U.S. prices kept falling, and they stayed low. By 2012, the British were paying nearly five times what American manufacturers paid for natural gas. In Japan, it was more than eight times as much. That's a competitive advantage that energy-intensive industries simply cannot afford to ignore, if they're going to remain globally competitive.

In September 2012, Goldman Sachs issued a report called "The US Energy Revolution: How Shale Could Ignite the US Growth Engine." In it, the asset-management arm of Goldman Sachs projected that the "potential for a sustainable US competitive advantage due to low domestic natural gas prices could allow the US economy to outperform other economies, with implications for US interest rates, the dollar and risk assets." The biggest beneficiaries would be steel, metals, chemicals, infrastructure, refining: industries that are massive users of energy, and industries in which the United States once dominated but have had to decamp for places with cheaper inputs.

Already, there are signs that once-shrinking industries are coming back to the United States, specifically for the energy advantage. Not that long ago, in 2008, there wasn't a single member of the American Chemistry Council who projected further investment in the United States – the petrochemical industry is a huge consumer of energy, and the United States just couldn't compete. By 2013, with U.S. natural gas selling for a third of the price of European rates and one-fifth of the price in Asia, the Chemistry Council was listing 110 new investment projects in the United States, worth nearly $80 billion.

The American economy once again rising rapidly in the world will have major geopolitical implications, of course. A weakened America, severely indebted to Chinese bondholders, is a vulnerable America. It's an America less able to defend its interests, and those of its allies, worldwide. It's a psychological shift, too: an end of the angst and weak morale that citizens feel when their country is shrinking in economic stature. We saw that during the Carter years, when Americans were plainly dispirited and Washington shrank in the face of world threats – that just happened to coincide with an energy price shock too. But it's also a psychological shift for everyone else: for U.S. allies, who need the confidence that they are backed by a ready and able America, and for U.S. enemies, who are emboldened by signs of American weakness.

But there is a more direct effect that the shale-gas revolution will have on American interests. Before shale gas began turning the energy world on its head, the United States wasn't in that much better a position, energy-wise, than Ukraine and Poland: because the U.S. economy needed to import billions of barrels of oil every year from outside the country, historically it had no choice but to patronize its OPEC enemies. So, even while the United States may have been outwardly committed to the cause of spreading freedom and democracy worldwide, it was actually sending hundreds of billions of dollars every year to the regimes that were working against it. Places like Saudi Arabia, Venezuela, and Russia. Places that used those billions of dollars to build up weapons, repress human rights, or even invade neighbouring countries, as Russia has done, or to sponsor terrorism, as Arab regimes do.

Saudi Arabia is one of the most odious regimes on the planet. That's no exaggeration: the human-rights watchdog Freedom House ranks Saudi Arabia in the bottom ten of the world's worst abusers of human rights. It's a dictatorship – a theocratic monarchy – where you'll still find slavery, where women are treated like men's property, and where sharia law metes out brutal punishments for offences that violate the Koran. Of course there's no democracy, no legitimate justice system, and no freedoms of any sort. But Saudi Arabia isn't just a threat to its own people: it's both the cause and sponsor of Islamist terrorists, like Osama bin Laden.

And yet Saudi Arabia is one of America's closest allies. Not because they share the same values or the same global interests. There are few countries in the world as diametrically opposed to American values as the Saudis are. The United States is pro-freedom, pro-democracy, pro-Israel; the Saudis are the polar opposite. But for Washington, there has simply been no alternative but to cozy up to Riyadh. The Saudis are the most dominant player in the oil market, and the biggest force within OPEC itself. The United States has had to put aside all its better judgement, all its foreign-policy ideals, just to ensure it remained on good terms with the Saudis. Because the Saudis had the power to severely sabotage the American economy: they did it in 1973, when they orchestrated an Arab oil embargo against the United States in retaliation for Washington's support of Israel during the Yom Kippur War.

After 9/11, it became even harder for America to bear this arrangement of enriching the Middle East dictators, sabotaging its own foreign-policy principles. And for the last decade, to its credit, Washington has been aggressively displacing Mideast imports with Canadian imports. Fortuitously, it so happened that Canada had just embarked on an energy revolution of its own. The Canadian oil sands – which like shale gas had a history of being largely uneconomical due to the complexity of separating thick, bituminous oil from sand – had become economically competitive, thanks to technological innovations (coupled with the rising price of oil). Canada was in a position to ship huge amounts of oil to the United States, enough to displace significant amounts of OPEC imports. But oil is mainly used for cars. Natural gas is a perfect fit for power plants, chemicals, and even transportation.

The technological promise of Canadian oil sands oil bumped up against a political problem: environmental extremism. The proposed Keystone XL pipeline – to crank the taps of Canadian oil, to finally get the United States off OPEC's conflict oil – became the lightning rod for environmental activists such as the Rockefeller Brothers Fund and Hollywood activist Darryl Hannah. They have convinced Washington to keep withholding approval for the pipeline to cross the U.S. border. And it's the same activists that have the potential to block the energy freedom promised by the shale-gas shift. In that Goldman Sachs report predicting that "shale energy could ignite the US growth engine," the bank lists the number one risk to the projected "energy revolution" as "litigation that curtails the practice of hydraulic fracturing." In other words, activist pressure, through the courts and elsewhere, aimed at pushing America and countries overseas away from fracking, permanently.

Where that would leave us, of course, is precisely where the environmentalists and peak-oil doomsayers have been saying we're headed all along. The primary political narrative about energy for the last two decades is that the world is steadily running out of fuel. That was the pretext for all the state-interventionist pressure away from carbon-based fuels and toward so-called alternative energy. The social engineers behind the Kyoto Accord and its successor, the Copenhagen Accord, were themselves motivated by other matters: they wanted the developed countries to phase out fossil fuels because they believed that carbon emissions were endangering the climate. But they also knew that you cannot persuade the voting public in North America and Europe entirely using those arguments: they needed to emphasize that fossil fuels were a "dead end" resource anyway, that the world was quickly running out of them, and they would become increasingly expensive to the point where they might financially cripple entire economies. If that was the future, after all, then the time to start switching to alternative energy was now, not waiting until it was too late. The "dead end" of fossil fuels was part of the sales pitch for carbon taxes – taxes on emitting carbon dioxide. Taxing CO_2 would lead to rationing. That was an environmental strategy, too, but in order to convince people that rationing carbon is necessary, you must first convince them that they are facing a scarce supply of carbon-emitting fuels. In 2003, *Time* magazine's Pulitzer Prize–winning reporters Donald Barlett and James Steele wrote an article entitled "The U.S. Is Running Out of Energy." They started this way:

> Natural gas is in scarce supply. Crude-oil production is winding down. The last nuclear power plant was ordered in July 1973. No meaningful alternative fuels exist. In short, Americans are heading toward their first major energy crunch since the 1970s. The early warning sign: a shortage of natural gas last winter sent home-heating bills spiraling upward. They are expected to keep rising. Higher prices are erasing jobs. The effects will ripple through the economy.

Americans, and Canadians and Europeans too, were so convinced that their fuel tank was running dry that they rationalized billions of dollars of investments in wind turbines and solar panels and geothermal projects. A mere decade ago, many bright minds were convinced that developed countries were heading for a colossal energy crunch: less energy, higher prices. We even began learning to live like fuel paupers. Every year, we were told to turn off all our electricity for an hour – for Earth Hour. It was a PR attempt to make it seem as if this North Korean–type existence was somehow noble. We would meet our cold, dark future with stoicism, even dignity.

It's not Barlett and Steele's fault that they were so spectacularly wrong. Almost no one foresaw the magnitude of the change that shale-gas fracking would bring (though there were more than a few optimists who could at least point to history as proof that technology consistently finds solutions to scarcity problems). But wrong they were: the United States is arguably more energy secure today than it has been in a century; the future of energy in America doesn't look like Earth Hour any more. It's not sitting in the cold and dark, accepting a wretched tomorrow with resignation. The future is filled with light and warmth and mobility and prosperity.

But that dark, cold future may well happen in some places, where they don't embrace the fracking revolution for political reasons. Places that refuse to exploit their own shale gas, and perhaps even refuse to consume shale gas, could well be facing a future of energy poverty – of prohibitively expensive fuel, whether that's oil, or gas, or even pricey solar or wind energy.

Energy poverty has an apt name. Running low on fuel means running low on everything else. The more expensive that heating your home, turning on your lights, and driving your car become, the less money you have for other things: leisure, education, furniture, even clothing, food, and shelter. That's why one of the most outspoken opponents to the anti-carbon environmentalist lobby in America is CORE, the Congress for Racial Equality. CORE was behind some of the most pivotal American civil rights campaigns in recent decades. Today, one of its biggest battles is against extreme environmentalists – the type who want to switch everyone off cheap fossil fuels and on to expensive renewable energy. It's a CORE issue because in the United States, it's black people who make up the largest ethnic group living in poverty, and CORE considers the environmentalist war against affordable energy a "war on the poor" – a war whose victims are largely black. The low-income Americans that CORE is worried about spend the highest proportion of their income on heating and gas. The U.S. Department of Health and Human Services calculated a few years ago that the poorest Americans face an "energy burden" – paying for gas and heat and electricity – that consumes, on average, half of their income.

"Environmentalists [are] . . . often speaking from the university, or they're speaking from a lifestyle in upper-class to a rich, privileged lifestyle that can afford to bear the costs" of expensive energy, Niger Innis, CORE's spokesman, has said. "The people that pay the price for the comfort of their intellectual exercise are poor people."

The Goldman Sachs report specifically projects that the "biggest impact" of shale gas in the United States "is likely

to be cheaper electric power generation." The biggest consumers of natural gas are electrical power stations, making
up as much as 34 per cent of the demand for natural gas.
Cheaper electrical power means all those impoverished
Americans spending half their income on energy suddenly
get a raise. Lower electricity bills mean they'll have more
money for the other basics of life. For some people, that
may mean the ability to move into a better part of town. Or
to save for college. Cheaper electricity isn't just a nice treat;
for some people, the difference between a $300-a-month
energy bill and a $100 energy bill can actually mean
opening up choices in life that they couldn't have before.

Certainly there are those who won't let up on their campaign to battle fossil fuels, no matter how cheap or plentiful
they may be, insisting that they are leading us toward
climate change, or global warming. They're largely the
same people fighting against shale-gas fracking. They have
a special hatred for fracking, because cheap, plentiful fossil
fuels actually represent a step backwards for their agenda
– a step further from their vision of a zero-carbon future.
The shale-gas bonanza undermines one of their most persuasive arguments for carbon rationing: that we're running
out of fossil fuels.

But if your main priority truly is reducing carbon emissions, you should actually celebrate fracking. The existence
of so much cheap gas, and the prospect of so many decades
of it, has naturally prompted a rapid and substantial shift
in the American energy economy away from other, less
optimal fuel types and toward cleaner, lower-emission

natural gas. Global-warming worriers can wring their hands about our fossil-fuel–based economies, urging us to spend billions more on solar and wind power. But economists have demonstrated that there is simply no way to feasibly replace all the power now generated by coal with wind turbines and solar farms.

By far the biggest advance in emissions reductions in the United States has been the spreading, countrywide re-orientation away from higher-emission power from coal and toward gas. A natural gas plant replacing a coal plant can produce the same amount of electricity but with emissions that are just a third of the level of that of the coal plant.

And that's just the reduction in emissions of carbon dioxide – a harmless, colourless, odourless emission. Gas plants are even better when it comes to real pollutants: sulphur emissions, particulate pollution, and other actual environmental toxins that are real and immediate and affect our urban quality of life, rather than prospective and speculative, like the theory of man-made global warming.

But whatever you think about carbon emissions, the vast America-wide switchover to more efficient natural gas is doing what dozens of international "climate summits" and "renewable strategies" could not do: it's reduced American CO_2 emissions in 2012 to levels not seen since 1994. The U.S. Energy Information Administration traced the decline to one primary factor: cheap natural gas. It reported that in 2012 "lower natural gas prices resulted in reduced levels of coal generation, and increased natural gas generation – a less carbon-intensive fuel for power generation, which shifted power generation from the most carbon-intensive fossil fuel (coal) to the least carbon-intensive fossil fuel (natural gas)." While environmentalist activists were holding "vigils for climate justice," politicians were giving away billions to unworkable alternative energy programs, and UN bureaucrats and lobbyists jetted to lavish conferences to draft meaningless global charters, it was the private energy industry that was busy developing and deploying technologies that actually reduced carbon emissions at a remarkable rate.

For the extreme eco-purists – the people who want the world off all fossil fuels, everywhere – even that kind of sensational progress isn't enough. The only grudging credit some of them will grant natural gas is to call it a "bridge fuel": a lower-emission alternative to oil and coal that can be used temporarily until their green alchemists develop that far-fetched fantasy fuel of the future. That's what the left-leaning Center for American Progress calls it: a "bridge fuel" to hold the U.S. economy over while we "conduct research on more efficient turbines, storage of renewable electricity, and other technologies that would generate no- or low-carbon energy."

Many of them won't even allow it that much though. "Natural gas isn't a bridge fuel – it's a gangplank," writes Josh Fox, director of the anti-fracking film *Gasland*. These are puritans who would make the perfect the enemy of the good – they'll block every fossil fuel, even one with a minimal environmental footprint, like natural gas, while they wait interminably for a miracle magical energy source that may very well never come.

But they are in the extreme minority. Even President Barack Obama, who is hostile to fossil fuels, has been willing to accept natural gas is a "transition fuel." It is more than that, of course. It's a real alternative to coal, a real reducer of carbon emissions (for global warming worriers), a reducer of pollution, a real energy source that is affordable and plentiful. If the true believers in a zero-carbon economy want to call it a "bridge fuel" while they wait for the arrival of cold fusion or dilithium crystals, then let them call it that. Lucky for them, for us, for the world, we've got so much natural gas that we can build a very, very long bridge to keep us going while we await their messiah fuel.

Those people are idealists. And there is a place in this world for idealism. But the true test of character isn't how moral they claim they would be in some hypothetical scenario. The true test of moral seriousness is how we make real decisions between imperfect choices.

Talking about what energy source might or might not be invented in forty years and then what might or might not be implemented and used, pretending that we can all get by with bicycles and windmills is childish – or at least unserious.

Ethics are about trade-offs. The trade-off for natural gas is western shale gas versus Russian Gazprom gas, Iranian ayatollah gas, or Qatari sharia gas. It's the choice between ethical energy and conflict energy.

A few years ago, I wrote a book that explored the same idea regarding the oil business. It was a defence of the Canadian oil sands, which were also being attacked by utopian idealists who insisted that the vast deposits of oil in Canada shouldn't be developed because we needed to move away from oil altogether. But that, too, was a false choice: the world wasn't about to give up oil in the foreseeable future. In fact, oil consumption is rising worldwide. The realistic choice isn't between oil and no oil; it's the choice between ethical oil from free, democratic, peaceful, and environmentally conscientious places like Canada, Norway, Britain, and the United States, and conflict oil from odious, repressive, belligerent, and environmentally reckless regimes, like Saudi Arabia, Russia, Venezuela, and Iran.

And it really doesn't matter if you're a conservative or a liberal, the ethical choice is clear. Actually, as I argued in my book *Ethical Oil*, the choice should be clearest of all to liberals. It's an irony that's lost on most liberals, of course, but the very things they have spent their lives fighting to promote, the very measures they have used to define progress, are the very same reasons we must favour ethical fuels over conflict fuels.

All oil burns the same. But OPEC oil is produced without any meaningful environmental protection; OPEC oil funds wars and terrorism; OPEC countries typically treat workers awfully; and OPEC countries uniformly violate basic human rights.

Environmental responsibility, peace, the treatment of workers, and human rights – that's the framework of the liberals, the same liberals who reflexively oppose things like oil sands oil from Canada and, increasingly, shale gas from fracking.

And just as Canadian ethical oil is superior to OPEC conflict oil on each of those four liberal values, so too is fracked shale gas ethically superior to the alternatives. We're not talking about money any more. We're talking about justice. Whether you're on the political left or the political right, this is the most ethical fuel you will find on the planet, at least until we find that magical mystery fuel. Ethical gas is like the "fair trade coffee" of the energy production world.

The alternative to producing our own energy, in places like the United States, Canada, and democratic Europe, is relying on imported energy. Before the shale-gas revolution came along, that generally meant relying on energy imported from some of the worst places in the world. The three biggest gas producers on the planet, controlling 60 per cent of the world's conventional gas reserves, are Russia, Iran, and Qatar. All the top countries for oil reserves, with the exception of Canada, are just as authoritarian and illiberal. They are what we'd call "conflict" countries.

And they're not particularly environmentally conscious, either. The energy giants of OPEC and gas OPEC have nothing like the environmental protections we take for granted in North America and Europe. Take Nigeria, with more than two thousand toxic oil spills that will never be remediated. In OPEC dictatorships like Saudi Arabia, the lack of a free press ensures that environmental disasters are never reported. On the Environmental Performance Index (EPI) compiled by Yale and Columbia universities, Russia's environmental ranking was 121st out of 163 nations. The Russians reportedly spill about five million tonnes of their oil every year – the equivalent of one Deepwater Horizon–sized spill every eight weeks – into their rivers and streams that flow into the fragile ecosystems of the Arctic Ocean.[4] Nigeria is even worse, with an EPI ranking of 153 out of 163. Saudi Arabia, ranking a dismal 99th on the EPI, covered up a gargantuan Persian Gulf oil spill – three times larger than BP's Deepwater Horizon spill – for seventeen years.

Liberal environmental activists from North America and Western Europe don't pay attention to the conflict nations that really are despoiling the earth. In a way, it would be pointless of them to truly care: Vladimir Putin and the Saudi King couldn't care less what Greenpeace thinks of their environmental record.

But buying OPEC oil and gas is environmental NIMBYism on the grandest scale, a form of pollution imperialism. It's so obvious that pollution protections in developed Western countries – the United States, Canada, Western Europe – are the strictest in the world that it's almost banal to point it out. And yet we're so bombarded by the incessant doomsday propaganda coming from environmentalists that we end up losing sight of that reality. But they have a deep financial interest in keeping us upset and worried: that's how they raise their funds, by alarming liberal consciences. And with so much clean water and air, and all the environmental protections, already in places like Canada and the United States, they have to find new reasons to make us worry. They need to spread panic about speculative and abstract fears: that we might be heating up the planet with our carbon dioxide; that science might one day find something dangerous about fracking.

For fracking, they've combined those unproven fears, making for a double-barrelled terror: natural gas fracking is poisoning our ecosystems, they tell us. And even if it isn't, it's overheating the atmosphere.

But of course, we already know that natural gas fracking isn't putting more CO_2 into the atmosphere – it's actually reducing the amount of previously emitted CO_2 at an astonishing rate, as it rapidly replaces coal as an electricity-generating fuel source. And the amount of CO_2 produced by all the natural gas burned for all reasons in the United States in 2011 – every molecule of gas burned in the world's largest economy – amounted to approximately 1,200 megatons (million tons).[5] That might sound big, but to put that into perspective, the amount of CO_2 released into the atmosphere by entirely natural causes is 770,000 megatons, more than 600 times as much as U.S. natural gas. If you're genuinely worried about CO_2 emissions, you've got bigger problems than natural gas, which amounts to the equivalent of 0.16 per cent of all the CO_2 belched out every year by forests, volcanoes, fires, oceans, and animal farts.

But shale gas isn't just the most ethical fuel choice for liberals panicked about CO_2 emissions; it advances all those other liberal values, too. When we support fuel from democratic, liberal countries, we support democracy and liberal values. Supporting conflict gas from Saudi Arabia supports repression and human-rights abuses against women and homosexuals and terror sponsorship. Supporting conflict gas from Iran supports a nuclear arms program designed to destroy Israel and to threaten and intimidate a long list of other countries in Asia and Europe. Supporting Russian conflict gas supports the intimidation and even occupation of former Soviet satellites and the systematic persecution of democrats and journalists.

Supporting shale gas in free, democratic countries means supporting gas produced by well-paid, often-unionized workers who operate in regulated, safe work environments. Supporting conflict gas from Saudi Arabia, Kuwait, and the United Arab Emirates (UAE) means supporting an industry that cruelly exploits workers. In the Gulf states, the bulk of the hard work of oil and gas production is outsourced to foreign indentured workers, often fellow Muslims, transient workers from Asia who have no civil rights, are forbidden by law from unionizing, typically have their passports seized upon arrival, and have their shoddy accommodations deducted from their meagre paycheques. Cities like Dubai look like gleaming jewels, but their labour practices are really just one step up from slavery – in the UAE, for every Emirati citizen, there are four foreign workers actually building the country. And it's not just men: throughout the Gulf millions of Asian women are often abused physically, not just economically.

Compare that to the American energy heartland, where two women serve as mayors: Annise Parker, of Houston, Texas, and Betsy Price, of Fort Worth, Texas. In Fort McMurray, Alberta, Canada's powerhouse industrial energy city, the mayor is Melissa Blake, and in Germany, the chancellor of the country is a woman, Angela Merkel. Women in positions of power have become largely unremarkable in the West, where women sit in legislatures, run major corporations, and sit on the Supreme Court.

Compare that to Saudi Arabia. It is against the law for a woman to be a mayor there. Or to vote. Or to drive. Or to get elective surgery without the permission of her husband/owner. Then there's the crime of adultery. Like the Islamic Republic of Iran, Saudi Arabia follows sharia law, and the punishment for "adultery" – the penalty, in other words, for being raped – is death by stoning.

It's not just women who are abused there; religious minorities are too. And, according to the U.S. State Department, in Iran alone, 4,000 to 6,000 gay men and women have been executed since the 1979 revolution simply for their sexual orientation.

This is the ethical choice we make when we shun energy from free countries – we are left instead to rely on energy from repressive and fanatical countries. It's easy for environmentalists to criticize and attack industry where we live, because the government doesn't throw them in jail for it. Not only that, whenever a newscast or newspaper article covers energy, it is virtually guaranteed they will actively seek out a dissenting quote from some anti-energy zealot – or several of them. And eco-activists can get away with protests that stymie development projects, chaining themselves to bulldozers or blocking railway lines. It's a paradox: the world's professional environmental activists prefer to condemn the gentlest energy sources in the world, rather than the bloodiest, precisely because these places are so gentle.

But whatever their complaints, the fact is that we face two choices when it comes to energy: we can get it from places that respect democracy, freedom, women's rights, minority rights, gay rights, and that implement strict protections for workers and the environment. Or we can get it from places that disregard all of those things. We can make an ethical choice, or we can make an unethical choice. And when we support gas from liberal, Western democracies, it means we deny support to the most heinous regimes on the planet. Who knows? Maybe one day scientists will invent that elusive wondrous power source that will satisfy both the anti-fossil-fuel zealots and the needs of our energy-dependent First World economies. But until then, any liberal with a conscience should be choosing ethical gas over conflict gas.

Chapter Two

AN OPEC FOR GAS

F racking might seem like a bright, inevitable future for plentiful, affordable natural gas.

But it's actually only one possible future. In places as diverse as France, Bulgaria, and Quebec, where fracking is illegal, the future of natural gas is something else: a future of imported dependence. And that suits most of the world's most powerful gas-producing nations just fine.

Remember that for some countries, such as Russia, gas is power. Not just electrical or transport power, but political power. These are places that had mapped out a future where natural gas was a central part of their economic sustainability and growth. Gargantuan sums of money are riding on the future of natural gas, trillions of dollars all told. There is geopolitical clout on the line. It would be naïve to think that those countries with a significant interest in one particular future path for gas won't do a great deal to make sure they get their way.

Iran has the world's largest gas reserves. Russia is close behind, at number two. And Qatar is number three.[1] These countries have dominated the world natural gas market for years. They also happen to be some pretty rough customers – Iran and Qatar are nasty dictatorships, and Russia has an authoritarian regime that seems to be regressing to a darker past. The United States produces a massive amount of natural gas every year – it's actually one of the world's most active producers. But it uses much of it up domestically. And its reserves are a fraction of the size of the Big Three gas giants. The United States has about 664 trillion cubic feet (tcf) in natural gas reserves.[2] Canada, another big energy producer by Western standards, has just 68 tcf. Now look at Iran: 1,186 tcf – more than triple U.S. reserves. Russia has 1,688 tcf. And Qatar has about 900 tcf. All the gas in Canada and the United States combined is still paltry compared to any one of those three gas giants.

Among them, Iran, Russia, and Qatar control 60 per cent of the world's natural gas reserves.[3] That's a market dominance they're not prepared to give up. So what do you do when you have the corner on a critical commodity market but you're facing an incredibly disruptive threat like fracking? A technology that has the potential to open up gas to a globalized, free-trading, fungible market? If you're a trio of insecure, bullying, regionally ambitious illiberal regimes whose futures rely on keeping gas supplies constrained, well, you team up to stop it.

And that's exactly what Iran, Russia, and Qatar are doing. They want to create an OPEC for gas. Remember what OPEC is: it's nothing more than a price-fixing cartel. OPEC countries meet regularly to blatantly collude on how to manipulate the world oil market. They decide how high they want the price of oil to go and how much supply they should release to the marketplace to ensure they get the price they agree on.

Until now, that kind of collusion wasn't as big a priority for natural gas. That's because the gas trade has never featured anything like the fluidity of the oil market. Oil is fungible – it's a commodity in the true sense of the word. Oil moves around the world through pipelines, like gas, but also by ship and by train. An oil importer looking for one million barrels of oil can order from Nigeria, or Venezuela, or Norway. That's why the biggest oil producers collude on oil prices: to limit competition in the market. OPEC was created as a response to American oil producers undercutting Middle Eastern producers. The sheiks and mullahs don't like competition. Fossil fuels are really their only source of economic revenue; they can't afford to chase competitive Western producers on price.

Gas is a much different story. There is some global movement of liquefied natural gas (LNG), but nothing on the scale of oil. The import and export of LNG requires specialized liquefaction facilities at the export terminal, and specialized re-gasification facilities at the import terminal. Oil just requires a receiving pipeline. And LNG is a relatively recent business: LNG shipping didn't even begin until 1959,[4] and there are still only about 365 LNG carriers active world-wide.[5] By comparison, there are more than 4,000 oil tankers.

Natural gas remains a largely landlocked trade. That means supply is constrained by geography – and politics. In the past, if Russia wanted to gouge Ukraine or Poland on gas rates, it didn't need to consult with Iran first. Ukraine and Poland were captive customers: if those customers didn't like the price Russia was charging, it didn't matter. There simply is no gas pipeline from Iran to Europe – although, not surprisingly, Iran is trying to build one.[6]

Unlike crude oil, there isn't a single, global market for natural gas. Everyone knows, roughly, the world price for oil – it's become a staple on nightly TV newscasts, the number having taken on a cosmic significance ever since the Arab oil boycott in 1973. There are some local variations in the price, and different regional sources of oil have unique names, like West Texas Intermediate, or North Sea Brent. But with a few minor wrinkles, a barrel of crude oil costs the same in New York, London, or Tokyo.

Not so with natural gas. There isn't a global price; in many cases, there isn't even a continental price. And so on the same day that natural gas sells in Poland for $14.90 per thousand cubic feet, it sells for $10.80 in Germany, $8.90 in the U.K., and around $4 in the United States, the birthplace of fracking.

Part of the explanation for this is benign – a result of expensive natural gas infrastructure, like pipelines and ships, slowly catching up to growing demand. Oil has simply been a bigger deal, for longer. These days, almost half[7] of the world's crude oil travels by tanker ship, and for very long distances, very cheaply. A single megaship like the *TI Oceania* can carry more than three million[8] barrels of oil, worth more than a quarter billion dollars at market prices – for context, just three of those ships could supply all of America's oil imports for a day. Fully one-third of all the merchant ships in the world are oil tankers, measured by cargo; over 4,300[9] oil tankers, with a massive capacity of 507 million[10] dead-weight tons, are in the world fleet. If there's a place somewhere in the world where an unusually high price exists for oil, it won't be long before some shipping company takes the opportunity for some arbitrage and ships in some cheap oil to fill the demand.

Compare that to natural gas, which is much more difficult to move by ship than oil is – the gas must be liquefied under great pressure and low temperatures, which requires an incredibly expensive processing facility; Australia's massive Wheatstone project, which will ship gas to Japan and Korea, is a staggering $29-billion construction[11] project. Imagine the scale of profiteering that's going on now by Russia and Qatar, to such a degree that investing $29 billion to provide competitive gas to Japan and Korea is still regarded as a profitable venture. And that $29 billion is just the first cost; the liquefied gas then has to be put in specialized LNG tankers. And once they're loaded, they can't just pull up anywhere; only another LNG facility on the other end can unload them too. There are just 350[12] of these LNG ships in the whole world – less than one-tenth the size of the oil tanker fleet – and while it's a growing business, only a third of the global[13] natural gas trade is by ship. So low-priced gas supplies in one country can't easily relieve high-priced demand in another.

According to the U.S. Federal Energy Regulatory Commission, the landed cost of an LNG[14] shipment – that is, the gas plus the cost to ship it there – ranges from a low of $3 per thousand cubic feet of gas in Lake Charles, Louisiana, to a high of $19.75 in Japan. China, India, Korea, Argentina, and Brazil are all over $14. Other than the United States, the only LNG markets where gas was less than $10 were the U.K. and Belgium.

There's an extra reason why Japan's LNG is the most expensive in the world: the country voluntarily decided to shut down all its nuclear power plants – not just the one in Fukushima that was damaged by the tsunami in 2011. More than 18,000 people died from the tidal wave, none from radiation from the reactor. Nonetheless, political pressure forced the closure of every[15] nuclear power plant in the country. To keep the lights on, Japan had to massively increase imports of other fuels, including coal and natural gas, causing electricity prices to spike by 30 per cent. Those high prices for imported energy pushed Japan into its first trade deficit in thirty-one years[16], mainly because of LNG costs. The country doubled its spending[17] on LNG imports, adding $20 billion to its energy costs. That's what life is like in a country depending on the Gazproms of the world – or at least one that chooses to be dependent. It was a self-inflicted case of energy poverty that helped propel Shinzo Abe, a pro-nuclear politician, to a sweeping victory as

Japan's prime minister in the 2012 election, on a promise to restart[18] the reactors.

When natural gas prices were high in the United States, before the fracking revolution kicked into high gear, LNG import facilities mushroomed along the U.S. Gulf Coast and New England (along with one in New Brunswick). But with domestic shale gas now cheaper than most imports, three of those LNG terminals have been re-engineered and authorized[19] for export, with more likely to follow. The idea of the United States – for decades an importer of fossil fuels, at the mercy of Arab sheiks – now exporting natural gas may seem unsettling, including to military and strategic thinkers who have longed worried about the security of the U.S. energy supply. But however unsettling it might be for Americans to see massive ships setting sail, taking U.S. gas to foreign shores, it surely is more unsettling to the Qatars and Russias of this world.

All U.S. LNG terminals touch the Atlantic Ocean, but a Canadian LNG terminal in Kitimat, British Columbia, is set to export that province's natural gas to customers in Asia. It's a huge facility, planned for five million metric tonnes of gas a year (mtpa), with a government licence to export twice as much if desired.[20] At 5 mtpa, that's $3.7 billion a year's worth of gas, at Japanese market prices[21] – each year.

Fracking has freed North America from expensive natural gas. And as the world's LNG fleet grows, Canadian and U.S. fracked gas could free Europe and Asia from the tyranny of the gas-OPEC monopoly, too. Because one thing Europe does have is plenty of access to open water. The LNG gas-tanker business is growing rapidly, and the revolution in shale gas is driving it faster every year. As of summer 2013, there were 31 liquefying and 80 gasifying LNG facilities on-stream, worldwide – 111 active facilities. But another 70 are right now under construction or in the planning phase. The world is about to see LNG infrastructure balloon by more than a third of its current size. And it's bound to keep growing as more countries, thanks to shale-gas fracking, look to export surplus natural gas. There are more than 100 LNG ships currently under construction worldwide, representing another one-third leap in carrier capacity.[22] More and more each day, the global trade in natural gas is moving toward the oil model. Even if Ukraine and Poland never develop their own shale gas industry (as unlikely as that is), they can break the stranglehold of Russia's Gazprom by building LNG import facilities and ordering tankers to sail up the Baltic Sea or through the Dardanelles to the Black Sea. And once a country becomes an LNG importer, it can build its own pipelines to neighbouring countries without open-water access. Gasification facilities in Poland and Ukraine could pipe gas to Belarus, Hungary, Austria, or the Czech Republic. The more natural gas becomes mobile, the more Gazprom's business model – its racket, really – is jeopardized. And the more Gazprom's racket is jeopardized, the more Moscow's regional power melts away.

That's why the big gas producers are so fearful of shale-gas production. It's also why they have an incentive to collude on supply and price. They're doing what they can to stop the fracking revolution – activities such as Russia's meddling in Bulgaria to instigate an anti-fracking ban there, and spreading misinformation and fear about fracking worldwide.

If you want a sense of how rattled these natural gas behemoths are about the threat that shale gas poses to their dominance, search for stories on fracking on Al Jazeera's news website. Al Jazeera is the international media group owned by the Emir of Qatar, Sheik Tamim bin Hamad Al Thani. Really, it's his royal messenger. Al Jazeera has tried in the past to pass itself off as independent, but even the staff don't really believe that: they've complained about Qatar's meddling – to promote the Sheik's interests – even in the channels broadcast outside Qatar.[23] The network has seen dozens of journalists, who evidently came on board believing the spin about editorial independence, resign in protest when Qatari bias compromised their reporting.[24] Nowhere is the bias more evident than in the way Al Jazeera covers fracking. Here are the headlines of some of their "reports":

"Fracked-off: Gas extraction 'causes quakes'"; "Irish critics warn against fracking"; "Fracking: A dehydrated UK, watered only by capitalism: When the UK's water infrastructure is already in severe drought, why is fracking even being considered?"; "Fracking ourselves to death in Pennsylvania"; "Fracking creates UK fissures"; "Fracking: A cure or a curse? To some it is a cure for an energy-hungry country, to others a flawed process that endangers people and the environment." It's one article after another about fracking, saturated with anti-fracking propaganda.[25] For an international news organization headquartered in Doha, Al Jazeera devotes a stunning amount of effort and space to the fracking issue, and it's pretty obvious why: exporting 4 trillion cubic feet of natural gas every year – more than double the entire annual gas consumption of France[26] – and with his entire economy dependent on energy sales, the Emir has everything to lose if his customers tap shale gas to become energy independent, or the natural gas market bursts wide open as countries everywhere suddenly become exporting competitors, thanks to gluts of shale gas.

But Al Jazeera on its own doesn't even need to mislead Americans and Europeans that fracking is a drought-exacerbating "flawed" "curse" that "endangers people and the environment." Not a lot of average Europeans and Americans are going to rely on Al Jazeera as the source for news and commentary, anyway. It's not necessary for Al Jazeera to succeed with its own propaganda efforts, as long as it provides paying work to all the anti-fracking writers and activists out there to contribute their own misinformation to Al Jazeera's website. That puts money in the pockets of a battalion of anti-fracking journalists who can then continue their work smearing the fracking industry in other media and financing their anti-fracking protest activities (naturally, Fox himself has been interviewed on anti-fracking–obsessed Al Jazeera's television programming as well). Qatar ranks as the richest country in the world, by GDP per capita;[27] it can easily afford to hand out indirect subsidies, paying anti-frackers to contribute to its "news" sites, to keep the anti-fracking campaign fires burning in the West.

But the big natural gas autocrats will never stop fracking everywhere, and so, with their control over 60 per cent of the world's gas reserves, they can at least try fixing the price, the way OPEC does with oil.

And they're already rolling out that very plan. In 2001, eleven of the world's big gas producers created the Gas Exporting Countries Forum (GECF).[28] That's an even larger group than the big three – Russia, Iran, and Qatar – and of course they control even more of the world's natural gas. Those three, plus Algeria, Egypt, Bolivia, Equatorial Guinea, Libya, Nigeria, Venezuela, and Trinidad and Tobago, control more than 70 per cent of the world's natural gas reserves, and 85 per cent of the world's LNG production. (By comparison, OPEC produces just 40 per cent[29] of the world's daily oil supply.) The GECF's officially stated mandate, originally, sounded almost harmless: "a gathering of the world's leading gas producers . . . set up as an international governmental organization with the objective to increase the level of coordination and strengthen the collaboration among Member countries." Coordination and strengthened collaboration – it almost makes it sound like some kind of natural gas United Nations. But look at that list of producers. Nigeria, Venezuela, Algeria, Equatorial Guinea? With Iran, Russia, and Qatar, it's a list of some of the most authoritarian, undemocratic regimes you'll find outside OPEC (most of them are OPEC members, too). And since the GECF began as the brainchild of Vladimir Putin, it was pretty clear from the start where this coordination and collaboration was heading.

And sure enough, just a year after its founding, Putin started talking about turning GECF into a "gas OPEC." In 2006, when Gazprom was battling its European customers over its price demands, deputy chairman Alexander Medvedev warned that Russia would build "an alliance of gas suppliers that will be more influential than OPEC."[30] At GECF meetings, Russia pushed to divide the global gas market into spheres of influence, with each producer monopolizing an assigned region.

Two years later, in 2008, Russia, Iran, and Qatar declared that they were going to form an OPEC-style gas cartel. Gazprom's chairman, Alexey Miller, announced the formation of a "Big Gas Troika."[31] You can probably guess that they didn't exactly test that name with focus groups. Big Gas Troika sounds like some oppressive Soviet pact – and really, that was the vision for it: a world-dominating bully that could intimidate the West using its control of the majority of global natural gas as a weapon. If the West got too shirty with Russia, or Iran, well, then, American allies in Europe and Asia would start to feel the pinch of rising gas prices. Some of them might even find their supplies cut off altogether. Just to drive home that point, Miller emphasized that the era of "cheap" hydrocarbons had come to an end.[32] Of course, he wished that were the case. That was the whole point of Big Gas Troika's plan.

With Iran's backing, by 2009, Russia's minister of energy declared that the GECF had gone from collaboration and strengthening relationships to becoming an even bigger Big Gas Troika. GECF had officially become a "fully fledged" "gas

OPEC," he announced – and naturally Russia was elected its secretary general.[33] "This is to show that member countries expect Russia to use its political weight to promote it." And he said, "We will work in sync to avoid overproduction of gas." In other words, restricting supply to inflate prices. Just like OPEC. As the Kremlin-controlled Voice of Russia broadcaster reported, GECF's key principle, according to Russia's energy minister, would be to "defend the interests of gas producing countries and LNG exporting nations."[34] Translation: tilt the balance away from consumers and toward producers.

This is the other future of natural gas, the non-fracking future. The illiberal, undemocratic bullies of the world carving up the globe into spheres of influence, constricting supply to raise prices, colluding to use gas as a political weapon.

The only thing that can genuinely dilute the power of the new Gas OPEC is the opening up of gas supplies in countries all around the world: in Poland, Ukraine, Bulgaria, France, Israel, and any other country sitting on top of significant gas deposits. The more shale gas that comes online, the less influence that GECF can have on gas prices – and the less they are able to manipulate the world's gas supply. The freedom that comes with gas independence is just waiting for so many countries to grasp. It's a future that can happen. But if the anti-fracking lobby succeeds in scaring more countries away from shale, the future of natural gas is going to look like the past. Only, with Russia's new gas cartel in the mix, rigging supply and prices, it will soon begin to look even worse.

Chapter Three

GAZPROM: HOW THE SHALE GAS REVOLUTION WEAKENS RUSSIA'S ENERGY MONOPOLY

T he workers inside know it as the "candle," so called for the glass pyramid that sits atop its roof. But the blocky concrete and glass of Gazprom's postmodern world headquarters in Moscow makes the tower look colder than a Siberian January. The four tall, sturdy, grey columns that soar, windowless, thirty-five floors up, on each corner are topped by smaller pyramids, like the spires of a medieval citadel. If Lex Luthor or some James Bond villain were in the market for a modern, vertical fortress, all dark and looming, he could do a lot worse than Gazprom's world headquarters. It even comes with that window-wrapped lair at the very summit, perfect for gazing out and cackling, fingers tented, plotting the subjugation of all the people below. Gazprom even had designs on erecting a new, even more imposing headquarters, in St. Petersburg – a

1,000-foot swooping flame of glass – and would have built it atop an ancient archaeological site, thanks to an illegal deal with city officials.[1] You could have ripped that story right from the pages of a comic book – if it had been a caped hero who had foiled the plan, instead of a local court, that is.

But outside appearances can be deceiving. Inside, Gazprom's headquarters is all garishness and kitsch. It has all the subtlety and class of a Donald Trump casino. A small fountain bubbles in the foyer, in the glow of a blue light – meant to represent a natural gas flame – but inexplicably cased in a box of steel-framed glass panels. Behind, giant flat-panel video screens run slideshows of volleyball players from Gazprom-sponsored Russian sports teams, meeting with Alexey Miller, chairman of the Gazprom Management Committee.

Miller is the man who dwells in the glass pyramid. It is here, Gazprom's website tells us, that "he holds the most important talks and festive ceremonies." And indeed, the penthouse looks quite like a mash-up between a Studio 54 lounge space and Khrushchev's cabinet room. There are two ways to enter Miller's perch in the sky: a circular steel staircase emerging from a G-shaped hole in the blue-carpeted floor – G for Grazprom – or a private elevator, wrapped on all sides in high-gloss mirror and stamped with the Gazprom logo. Subtlety is not exactly a company forte.

Massive balls of wire implanted with glowing bulbs, like disco balls wrapped in Christmas lights, hang over the leather swivel chairs in Miller's waiting area; two black flat video panels hang over the circular boardroom table, surrounded by a dozen black leather meeting chairs. Overhead, a massive chrome and glass light fixture hovers, like a flying saucer from a low-budget fifties sci-fi movie. Step out onto the balcony on any side and take in the sprawl of southwest Moscow: an endless grid of bleak, grey Soviet-era apartment blocks and factories. For thousands of miles beyond are Gazprom's subjects – er, customers.

If you wanted to design a building that suitably embodied Gazprom's very nature – its monopolistic dominance, its ostentation, its forbidding nature, its state-run, Kremlin-dominated, crony-capitalist hybridism – you'd probably come up with something that looked a lot like the "candle."

And as you can tell by Chairman Miller's garish (and unironic) sky-lair, Gazprom isn't particularly self-conscious about showcasing its overbearing power. As it happens, Miller himself was installed as chairman by the newly sworn-in president, Vladimir Putin, in 2000, as a blatant move to marry Gazprom's gas business with Putin's new, authoritarian, and revanchist program. Critics complained that Miller possessed "utter lack of knowledge of Russia's oil and gas industry," but he was a Putin ally from way back, which is what mattered more to the Kremlin chief.[2]

The Soviet-era spirit isn't just shot through Gazprom's architectural aesthetic: it lives on in its operational strategy. When Gazprom carves up the European market for itself, it insists on retaining the old Warsaw Pact distinctions: there's "Europe" – Turkey, Italy, France, and so on – and then there's the "CIS," the Commonwealth of Independent

States.[3] That would be Russia, Ukraine, Belarus, Armenia, Kazakhstan, and a list of other former Soviet-run Communist puppets. To Gazprom, they still look like the same Russian-dominated Eastern-bloc empire, only one now ruled with a control over energy, rather than military might.

And Gazprom doesn't hide the fact that those former U.S.S.R. vassal states are treated differently. Right on Gazprom's "question and answers" website, the company presents a chart showing the different gas prices it charges different kinds of customers. In 2012, the average price the company charged Russians for natural gas was 84 rubles per thousand cubic feet. For the "CIS and Baltic States" – the Baltic States being former Soviet satellites Estonia, Latvia, and Lithuania – the average price Gazprom shows it charged in 2012 was 227, more than twice as high.[4] The company chalks that up to the added cost of exporting gas outside Russia's borders, which is certainly a consideration – but it's also true that many of Gazprom's production fields are actually inside those very former Soviet states. Gazprom produces gas in Uzbekistan, Kazakhstan, Tajikistan, and Kyrgyzstan.[5]

In the real world, this is not how commodities are normally supposed to work. No freely traded goods have vastly different pricing structures all within the same basic region. In a country that respected and implemented free trade treaties, as the United States, Canada, and the European Union have, for example, this kind of discriminatory pricing would be illegal. What Gazprom and Russia prefer, though, is unfree trade – the ability to use price as an economic cudgel to contain power within the region. That's why Gazprom continues to sign its import markets to twenty-five-year-long gas contracts, as it recently did with Kyrgyzstan[6] (and tried to do with Turkey,[7] while failing to get China to sign a thirty-year deal).[8] You'd have to be foolish, or under duress (as Kyrgyzstan likely was), to do such a thing: as shale gas revolutionizes the world gas supply, gas can only get cheaper than ever. But, naturally, Gazprom wants to freeze its regional supply dominance in amber for as long as it possibly can.

This isn't just a theory. Gazprom and Russia are blatant about the use of Gazprom as a political instrument. Putin has categorized Gazprom as a "strategic company," a pillar in his re-creation of Russian clout (Putin's Ph.D. thesis at the St. Petersburg Mining Institute argued that Russia's strength would rely on creating national energy champions).[9] But that has also allowed him to pass laws maintaining that its "strategic" value means he isn't obligated to disclose any of its operational details to the European Commission, which has tried investigating Gazprom for unfair business practices. Under Putin, it is now illegal for Gazprom to "trade with or divulge information to foreign entities without first receiving permission from the government," the *Moscow Times* reported in 2012. Putin has given himself veto power over all of Gazprom's international dealings, as if the distinction between his gas titan and the Kremlin itself weren't already blurry enough. By designating Gazprom as a "strategic" asset, Putin can treat it as if it were a jet-fighter or other arms manufacturer – which, given his practice of using it to subdue other countries, has a certain perverse logic to it.

In recent years, both Ukraine and Belarus have groaned under the crushingly high Gazprom rates they were paying, to the point where their economies were in jeopardy. The burden was felt not just by energy-dependent industries, but by ordinary homeowners too. At the time, the Kremlin was pressing its former Soviet satellites to sign on to a Moscow-led trade deal that the vice chairwoman of the National Endowment for Democracy, Judy Shelton, has described as Russia's attempt "to reassert not only economic but political dominance over the area that it formerly controlled" and a "step backward in terms of individual freedom" for the former Soviet countries.[10] Moscow offered Ukraine and Belarus a deal: sign on to the customs union and get a break on those extortionate gas rates. Belarus "caved," as the German paper *Der Spiegel* describes it, and now pays $5.24 per thousand cubic feet of Gazprom natural gas.[11] Ukraine, refusing to give in, is stuck paying more than twice that much: $12. Ukraine is hoping instead it can get shale-gas and LNG deals in place quickly enough to salvage its economy without succumbing to Moscow's extortion racket.[12] When Russia put Moldova, another former Soviet colony, in the same bind, threatening to keep prices high after Moldova expressed interest in joining the Western Europe–backed Energy Community, a bulwark against Russian energy domination, the European Commissioner of Energy, Gunther Oettinger, called Russia's behaviour "pure blackmail."[13]

This is the reality for so many former Soviet colonies that once lived in the shadow of the hammer and sickle: they now live in the shadow of Gazprom's fortress-like "candle," forced to choose between independence from the Kremlin and affordable energy for their citizens and industry. Lithuania has petitioned the EU to investigate Gazprom for unfairly inflating gas prices in retaliation for Lithuania's decision to break up a Gazprom utility monopoly in that country and deregulate energy markets.[14] Poland, too, has tried seeking redress over unfair pricing by Gazprom. Ukraine is stuck paying for overpriced gas it may never even use – they have to buy a fixed amount of gas, at a fixed price, in a long-term contract – though it could have gotten out of the deal had it been willing to hand over control of its pipeline network to Gazprom. Moldova has been warned that if it wants its Gazprom supplies to remain affordable, it must decline opportunities for European integration and remain, instead, locked into Russia's economic orbit.[15]

No private corporation would or could act this way. For all the ire directed by activists at ExxonMobil or BP, those companies are global Boy Scouts when you consider the breathtakingly brazen manipulation of sovereign governments that Gazprom employs in order to serve its masters in the Kremlin. This isn't corporate deal-making: it's like negotiating with the Red Army. And these are particularly wrenching dilemmas for countries that have strived for so long to liberate themselves from the Kremlin's iron grip. As one analyst put it in a Polish energy report, "Without energy independence, political independence from Moscow is hollow. There must be Eastern European solidarity on energy security if Russia's influence is to be curtailed."[16]

Gazprom would rather its retail end-users didn't think of it as a tyrant, but it can't exactly stop behaving like one, not when it's under direct orders from the Kremlin to strong-arm Russia's neighbours. So it's tried at times to at least polish its public image. But it's awfully tough to come up with ways to put a friendly face on an institution that is so explicitly designed to be overpowering and threatening.

Gazprom's U.K. communications office, for one, has been working to freshen up the company's communication efforts, to make it look "more like . . . a lively technology start up, not a bureaucratic state juggernaut."[17] The CEO of Gazprom's Marketing and Trading arm, Vitaly Vasiliev, has "hammered out the message . . . that Gazprom is an ordinary business that simply wants to deliver good customer service." There's even a Gazprom theme song, complete with a music video, available on YouTube,[18] to persuade customers that this faceless, merciless state-run conglomerate is also full of neighbourliness and good cheer.

"Don't bother trying, you'll never ever find a surer friend than Gazprom; We're giving people warmth and light, for office and for home," booms the singer, Vladimir Tumayev, who also composed the song and ran a Gazprom subsidiary, when he wasn't busy writing and performing passionate tributes to his corporate masters. "We're renowned for our deeds, the world over" goes another line – which is technically true, since the song doesn't actually get specific about whether it means good deeds or malicious ones.

But even in song, Gazprom's backward values reveal themselves. As the video flashes images of men working on high-pressure, highly combustible gas rigs, the chorus instructs: "Let's drink to you, let's drink to us, let's drink to all the Russian gas, that it never comes to an end, though it's so hard to obtain." You won't find too many lively technology start-ups advising the mixture of alcohol and operating heavy equipment while pumping substantial volumes of inflammable gas.

No matter how it tries, the company simply can't suppress its Soviet DNA. In 2012, Alexander Medvedev, Gazprom's deputy chairman and head of its export division, might as well have been channelling Khrushchev when he not so subtly threatened Europe over the European Commission's attempt to investigate Gazprom over its unfair business practices.[19] "Frankly, this makes you wonder if [the European Union] wants to receive Russian gas in growing volumes," Medvedev said, perhaps snickering at the cleverness of his implication: Mess with our business model, and it may be an awfully chilly winter for you. No PR whiz is going to be able to spin positive publicity on that kind of Cold War ultimatum.

It isn't just its buyers from the former U.S.S.R. that Gazprom puts the screws to. It bullies those former Soviet states that it buys natural gas from, too. Gazprom's reputation is as a gigantic monopolistic seller, but it throws around its weight as a monopsonistic buyer to force down prices from Central Asian countries with national natural gas reserves. Take Turkmenistan, a Gazprom supplier that BP now says has the fourth-largest gas reserves in the world, behind Russia, Iran, and Qatar. After independent audits revealed that Turkmenistan's natural gas fields contained 925.24 trillion cubic feet of gas, Gazprom began a misinformation campaign to scare other investors away, to keep the Turkmen supplies from becoming too attractive to other competitors. "I don't think there are reasons for making such statements [about gas potential], given that no real audit has been conducted, or a report presented on its results," Alexander Medvedev told a Russian television interviewer – despite the fact that there had already been at least two independent audits. "Of course, there is gas in place in Turkmenistan. But the new gas fields are not easy to handle and are in need of sophisticated extraction know-how."[20] Turkmenistan had announced plans to pursue "energy cooperation with all interested parties . . . and the diversification of routes to supply its energy resources to world markets."

The idea that a former Soviet state with massive gas reserves might be welcoming other investors to come in and produce natural gas, and that Turkmen gas might end up competing with Gazprom's supplies on world markets, is naturally intolerable to Gazprom, which will do whatever it can to dissuade any other interested parties from setting up shop, by understating the potential of Turkmenistan's gas fields and insisting they're not that easily tapped, anyway. Turkmenistan obviously knows it, too. After

Medvedev's comments, the country's foreign ministry issued a statement denouncing his remarks as "yet another clumsy attempt to distort the real situation regarding Turkmenistan's resource potential, in particular its gas reserves." But then, Gazprom has been messing with Turkmenistan's supplies for years – just because it could. In 2009, Gazprom unilaterally turned off the taps of gas supplies coming from Turkmenistan, with only a few hours' notice. That actually caused explosions along the Russian-bound pipeline, due to the sudden pressure drop. The explosions cut off Turkmenistan's supply route to Russia for nine months. By the time exports were up and running again, Turkmenistan had been forced into a weaker position by Gazprom's bully tactics: it would now have to sell gas to Gazprom at just two-thirds what it was getting before Gazprom unleashed its dangerous mischief.[21]

Gazprom typically and forcefully defends its massive monopolistic market power, as most monopolies do, by claiming it leads to price "stability" and "fairness" in the market, while maintaining that the high cost of drilling and pipeline infrastructure can be borne only by an immense company with the security that comes with a huge share of the market. Its entire shtick is that energy supplies need to be well-funded, predictable, stable, and secure. In reality, as Gazprom's buyers have discovered, when they find their supplies suddenly cut off in the middle of winter, or when they are forced to pay extortion-rate gas prices because they refuse to buckle to Moscow's political control, all that talk of stability is a one-way street. Even the president of the European Union, José Manuel Barroso, has said that Russia's behaviour was the "first time I saw when [gas] agreements were not respected. . . . Gas coming from Russia is not secure."[22] And it's no different on the supply side, where Gazprom's version of stability and security and fairness can mean Gazprom deliberately sabotaging your country's pipeline infrastructure, just to wring out a gas discount.

Of course, the most useful way to achieve security and stability in the energy market is to unlock as many gas reserves as possible, keeping prices low and giving the market plenty of competition. But the one thing that is most capable of opening up the natural gas market is the one thing that Gazprom is dead set against allowing: the widespread fracking of shale gas.

Gazprom executives sometimes seem so flustered by the prospect of a European shale-gas bonanza loosening their stranglehold over the market that they occasionally contradict themselves in the process. Sometimes they'll dismiss the idea that shale gas could possibly lower prices at all and insist it poses no threat to Gazprom's conventional-gas market share. One executive told Reuters in 2012 that the company is "thrilled" that Poland is starting to develop shale gas. It would cost a lot more than the Polish think it will, he explained, and presumably teach the country a lesson about taking Gazprom's prices for granted. "Look, we do not believe in this myth of shale gas, that it is cheap gas," said Sergei Komlev, Gazprom's head of contract structuring. "It is not true."[23] (In fact, Gazprom's own political shenanigans and corruption have a way of significantly driving up its production costs – such as deliberately routing pipelines so they avoid certain markets, as a way of restricting supply, as in Ukraine.)[24]

But at other times, Komlev doesn't seem nearly so "thrilled" about the prospect of shale gas taking off in Europe. In a meeting of the Gas Exporting Countries Forum in 2011, Komlev offered up a PowerPoint presentation to his fellow gas producers. In it, he presented a list of handicaps that "will retard shale gas development outside [the] US" – which included "economic, regulatory and political barriers."[25] The slide said "fortunately" shale gas might not make a significant dent in the European market for a decade or more. "Fortunately" for Gazprom – not so much for their consumers.

Komlev does, however, have the nerve to publicly complain that the push toward shale-gas fracking is part of a "political agenda" in Europe, presumably to weaken Russia. It takes a certain amount of chutzpah to say one thing about welcoming shale-gas competition when a Reuters reporter is taking notes, but to turn around and tell your fellow gas barons that you're grateful for all the barriers in place that will keep shale gas from making energy more affordable in Europe. But it's a whole other level of gall to complain that European countries have a "political agenda" simply because they're looking for ways to break free from a Russian gas behemoth that has been carefully structured to become a blunt weapon for Vladimir Putin's foreign policy.

In the meantime, Gazprom is doing everything it can to defame fracking, to stave off the inevitable European shale-gas revolution as long as possible by planting doubts through propaganda and political agitation. Gazprom hired the U.S.-based consultancy Pace Global, which has begun issuing negative reports on the future of shale gas (an example: "Shale Gas: The Numbers vs. The Hype") aiming to show that the affordability of shale gas is only temporary and that prices are bound to rise.[26] Both Gazprom and the Russian government are now reportedly funding anti-fracking environmentalist groups[27] though hard evidence, as with any Cold War–style intrigue, is foggy. The Kremlin-controlled international English-language all-news channel, Russia Today (RT), gave loving interviews[28] to Josh Fox about his anti-fracking film *Gasland*, but in fairness, they were not any more slobberingly positive than those on all-news TV channels in the West.

But Gazprom has begun its own campaign to sow environmental fears about fracking. Gazprom's Alexander Medvedev, for instance, has taken to pushing the myth about drinking-water contamination. "Not every housewife is aware of the environmental consequences of the use of shale gas. . . . I don't know who would take the risk of endangering drinking water reservoirs," he told a press conference in London in 2010.[29] "There are shale gas reserves in Europe, but I honestly don't think anybody would launch themselves into production using existing techniques. . . . Even the French would never agree with the replacement of their drinking water with wine." In late 2011, Gazprom's board of directors took the peculiar step of issuing a press release rehashing baseless anti-fracking fear mongering: "The production of shale gas is associated with significant environmental risks, in particular the hazard of surface and underground water contamination with chemicals applied in the production process," it read.[30]

Fracking simply does not contaminate groundwater – let alone pose "significant environmental risks" of any sort – and the Gazprom board and Medvedev certainly know it. We can be sure they do because Gazprom and other Russian energy companies are fracking in Russia. In fact, as Bloomberg business news reported back in 2012 – just a few months after the Kremlin-controlled Gazprom board issued that bizarre anti-fracking leaflet – Putin is betting on Russia's Bazhen Shale Formation to "someday match the explosive growth of North Dakota's Bakken shale discovery."[31] He's even offering tax breaks to entice more and faster drilling, to kick off a Russian shale revolution. Russia's former deputy prime minister, Igor Sechin, has predicted that unconventional production, such as shale drilling, will become a huge part of Russia's total fossil-fuel output. You see, Vladimir Putin isn't anti-fracking the same way that New York artistes and Hollywood celebrities are anti-fracking. He doesn't want to eliminate fracking entirely. He just wants to make sure that the only fracking being done in Europe, the only gas production done in Europe, is being done by Russian companies like Gazprom. Just as in the Cold War, Russia's interests are simple: unrivalled domination.

Chapter Four

CRITICISMS OF FRACKING: CONTAMINATED GROUNDWATER

All the anti-fracking hype is designed to make you believe that the U.S. government has been asleep at the switch when it comes to monitoring environmental safety. The activists want you to believe that a film director named Josh Fox can grab a video camera and, within a few months of driving around the country, easily expose a catalogue of hazards that all the experienced and educated scientists at the U.S. Environmental Protection Agency (EPA), not to mention all the state regulators, missed.

Fox even implies that President Obama has been naïvely misled on the issue by the dastardly oil and gas industry. In July 2013, Fox wrote an open letter to Obama, in which he reminds the president how frequently he's met with industry representatives. He implores the president to meet with him and seven families who "have all had their lives ruined" by fracking.[1] "We seek to discuss with you the dark side of fracking, a perspective that has not yet been presented to you with adequate weight or emphasis," he writes. Of course, if anyone knows just how informed the president of the United States is, it must be a crusading New York City filmmaker.

But the Obama administration has proven itself to be no booster of the fossil-fuel industry, and under Obama, the EPA has been no sleeping watchdog. In reality, they have been active and invasive, particularly when it comes to fossil fuels. Obama has been firmly against opening up Alaska's Arctic National Wildlife Refuge (ANWR) to drilling. And the EPA has shut down other drilling in Alaska because of fears over environmental impacts.[2]

After the 2010 BP Deepwater Horizon oil spill, Obama imposed a total, five-month-long moratorium on all deepwater offshore drilling across the U.S.'s entire Outer Continental Shelf, which covers the Gulf of Mexico as well as the Atlantic coast. That may have seemed prudent at the time, when the spill was still active and there were still so many questions about the safety of deepwater offshore drilling – despite the severe impact the ban was having on coastal communities that relied on the oil business. But even after the moratorium officially ended at the end of October 2010, the administration continued to refuse to grant permits to drillers so they could start up again. Another month lapsed, then another, and another. It wasn't until another four months had gone by, turning a five-month ban into a nine-month ban, that the federal administration finally granted a permit to Noble Energy to allow it to restart drilling.[3] Obama was in no hurry to help the oil industry get back to work.

Meanwhile, Obama has effectively put the coal industry on notice that he intends to put them out of business. Speaking to reporters in 2008, he described how he aimed to put in an environmental regime that would penalize heavier emissions so greatly, with limits being continually "ratcheted down," that if any company built a coal-powered plant, "it would bankrupt them."[4] Daniel P. Schrag, a scientific adviser to Obama, told the *New York Times* that "a war on coal is exactly what's needed" to meet climate-change goals.[5] And of course this president has declined to approve the Keystone XL pipeline, which would carry 800,000 barrels a day of Canadian oil to Texas refineries, despite the fact that his own State Department has determined that the pipeline will have "no material impact" on U.S. greenhouse gas emissions[6] and could find no environmental basis on which to reject the pipeline.[7]

The track record of Obama and his EPA, in other words, is one of acute, often even baseless, precaution. Which is important to keep in mind when you read the EPA's definitive review of fracking and its potential for contaminating groundwater. That is, they have found no proven cases of fracking-related contamination. Exactly zero. Not a single one, anywhere, ever.

"In no case have we made a definitive determination that the fracking process has caused chemicals to enter groundwater," Lisa Jackson, then head of the EPA, told a reporter in 2012. Now, you'd be hard-pressed to find a head of the EPA who was as enthusiastic an environmentalist as Jackson. She commonly promoted the concept of "environmental justice," which is based on the belief that poor people and minorities are most victimized by pollution and other environmental blights, and that additional regulation is necessary to ensure they receive the added protection they lack. Roughly, it puts environmental regulation on a footing with civil rights.

Jackson established a list of "Seven Priorities" for the future of the EPA, the first of which was "Taking Action on Climate Change," followed by "Improving Air Quality," "Assuring the Safety of Chemicals," and "Protecting America's Waters."[8] But there was also "Expanding the Conversation on Environmentalism and Working for Environmental Justice." Under Jackson's watch (she stepped down at the end of 2012), the EPA tightened gas-mileage standards for vehicles, ordered that gasoline must have a 50 per cent larger ethanol component, imposed more stringent ambient air-quality standards, and tightened regulations on coal- and oil-fired power plants.[9]

In advance of the Copenhagen climate summit – the would-be successor to the Kyoto Accord – Jackson had the EPA rule that carbon dioxide, the gas that we exhale countless times a day and that nourishes plant life on this planet – was effectively a pollutant requiring regulation, lumping it into a regulatory category along with methane, nitrous oxide, hydrofluorocarbons, perfluorocarbons, and sulphur hexafluoride, which the EPA ruled "in the atmosphere threaten the public health and welfare of current and future generations."[10]

And in 2012, the man whom Jackson had put in charge of the oil-rich South and Southwest regions of the United States was forced to resign after he said publicly that the EPA's strategy for regulation was like the Ancient Romans, who would immediately "crucify" a few people after invading a Mediterranean village, to scare the population and make them easier to manage afterward.[11] In Lisa Jackson's EPA, he said, "you make examples out of people who are, in this case, not complying with the law . . . and you hit them as hard as you can" to act as a "deterrent" to other companies in the same region.

This is the same crusading, crucifying Lisa Jackson–led EPA that had looked into all those reports of fracking-related groundwater contamination that we hear so much about in the media and from anti-fracking propaganda – and found that there was no evidence of it. Zero.

And the EPA has been on top of this issue for years, long before Josh Fox and his fashionable anti-fracking celebrity movement came on the scene. In 2004, the EPA released a study representing four years' worth of the agency's research into the safety and environmental effects of fracking.[12] It "reviewed incidents of drinking water well contamination believed to be associated with hydraulic fracturing and found no confirmed cases that are linked to fracturing fluid injection into coalbed methane wells or subsequent underground movement of fracturing fluids." It concluded, "The injection of hydraulic fracturing fluids into coalbed methane wells poses little or no threat to USDWs [under-ground sources of drinking water]." The study was suffi-ciently exhaustive, the EPA determined, that it did "not justify additional study at this time." The EPA's long-term research and scientific evidence is the sort of thing that President Obama would rely upon in continuing to allow fracking. When Josh Fox writes to the president insisting that his arsenal of anti-fracking stories have yet to be pre-sented to the president "with adequate weight or emphasis," it is because the science, the actual evidence, proves that all those fracking-contamination claims are mistaken – or that a filmmaker is looking for some publicity.

And it isn't just the EPA proving it. On the state level, too, over and over again, these tales of contaminated groundwa-ter have been found to simply have nothing to do with fracking. When Alabama regulators reviewed fracking activity in their state, they came up with the same result as the EPA: "There have been no documented cases of drink-ing water contamination that have resulted from hydraulic fracturing operations to stimulate oil and gas wells in the State of Alabama."[3] Researchers came up with the same goose egg in Alaska: "There have been no verified cases of harm to ground water in the State of Alaska as a result of hydraulic fracturing." And in Colorado, "no verified instance of harm to groundwater caused by hydraulic frac-turing." And "no instances" were identified in Indiana "that harm to groundwater has ever been found to be the result of hydraulic fracturing." Kentucky looked into complaints from landowners about contaminated groundwater but the results were predictable. "In Kentucky, there have been alleged contaminations from citizen complaints but nothing that can be substantiated." In Louisiana, regulators, too, are "unaware of any instance of harm to groundwater . . . caused by the practice of hydraulic fracturing."

Fracking has been going on in Michigan for many years; there are thousands of fracked wells in that state. If fracking really did contaminate groundwater, even occasionally, it would surely have happened in Michigan. But investigations there found "there is no indication that hydraulic fracturing has ever caused damage to ground water or other resources in Michigan." In fact, by 2009 when they reported that, Michigan's Office of Geological Survey said it had never even received a single complaint or heard a single allegation that fracking had affected groundwater "in any way."

In Oklahoma, they found evidence of groundwater contamination – from conventional oil and gas projects, that is. But from fracking? None. Despite the fact that "tens of thousands of hydraulic fracturing operations have been conducted in the state in the last 60 years," they reported. In Tennessee: "No reports of well damage due to fracking." In Texas: "Though hydraulic fracturing has been used for over 60 years in Texas . . . records do not reflect a single documented surface or groundwater contamination case associated with hydraulic fracturing." Drillers have been fracking for oil in South Dakota since the fifties, and for gas since 1970, and still the state reports "no documented case of water well or aquifer damage by the fracking of oil or gas wells." Same deal with their neighbours to the west: "No documented cases of groundwater contamination from fracture stimulations in Wyoming."

The Ground Water Protection Council, a non-profit organization whose membership consists of state-level groundwater regulators and whose very purpose is to "promote the protection and conservation of ground water resources for all beneficial uses, recognizing ground water as a critical component of the ecosystem," issued a report in 2011 that reviewed fracking in Texas and Ohio. The study covered sixteen years of activity, during which more than 16,000 horizontal hydraulic-fracking shale-gas wells were completed in Texas alone. In neither state did regulators identify "a single groundwater contamination incident resulting from site preparation, drilling, well construction, completion, hydraulic fracturing stimulation, or production operations at any of these horizontal shale gas wells."[14]

Not only have regulatory investigations everywhere across the United States found not a single drop of drinking water contaminated by fracking, but it isn't actually physically possible for something like that to happen. Why? Because in not one single case does a hydraulic fracture even come near the water table.

See, all of this fracturing is happening at nearly a mile, or deeper, below the earth – that's where the shale gas is. Water wells don't go nearly that deep. Typically a well goes down several dozen feet, or maybe even a couple of hundred feet if the water table is exceptionally deep. America's biggest hand-dug well, the Big Well in Greensburg, Kansas, dug in 1887, goes down 109 feet;[15] the Well of Joseph in Cairo's Citadel, in the Egyptian desert, goes down 280 feet. Those are deep wells, because they're built over deep water tables. Water aquifers are often deeper: they average around 500 feet below the ground.[16] But fracking? That happens thousands of feet below the surface – typically between 6,000 and 10,000 feet underground.[17]

For the gas or the fracking fluid to get into the water table, or even an aquifer, from that kind of depth, they would have to pass upward through millions of tons of rock – like passing through a mountain. In the Barnett Shale, for instance, even the shallowest fractures are roughly a mile below the surface – thousands of feet below any aquifer or water table.[18]

These facts have been on the record far longer than the media and activists had even heard of the term "fracking." In 1995, the EPA under the Clinton administration – who were no slouches, either, when it came to environmental restrictions – declared that "there is no evidence that the hydraulic fracturing . . . has resulted in any contamination or endangerment of underground sources of drinking water (USDW)."[19] The EPA had been studying fracking in Alabama as far back as 1989. "Moreover, given the horizontal and vertical distance between the drinking water well and the closest methane gas production wells, the possibility of contamination or endangerment of USDWs in the area is extremely remote." That was Carol Browner writing, the environmentalist lawyer who served as EPA administrator under Bill Clinton and later became the director of the

White House Office of Energy and Climate Change Policy under the Obama administration.

That doesn't stop anti-fracking agitators from trying to scare people into believing that it's not only possible but that it's actually happening and happening frequently. In his movie *Gasland Part II*, Fox cites a Cornell professor named Anthony Ingraffea – an anti-fracking activist himself[20] – who supposedly cites industry documentation revealing that 60 per cent of shale-gas wells will "fail" – leading the audience to believe that there will be some kind of leak that leads to the gas or fracking fluid escaping the casing around the hole and getting into the water supply after all. But the documents that Ingraffea and Fox are relying on actually have nothing to do with shale gas. The figure is related to SCP – sustained casing pressure – over a thirty-year lifespan of wells "in the outer continental shelf (OCS) area of the Gulf of Mexico, grouped by age of the wells."[21] The statistics from the United States Mineral Management Service, which are what Ingraffea and Fox are misrepresenting, have to do with wells in the ocean, and specifically "do not include wells in state waters or land locations."

But once Josh Fox says it, his celebrity fans repeat it, and then it becomes yet another talking point in the anti-fracking propaganda machine. The *New York Times* recently featured a letter from Yoko Ono, representing her group Artists Against Fracking, in which she repeated the lie: "Industry documents show that 6 percent of the wells leak immediately and that 60 percent leak over time, poisoning drinking water and putting the powerful greenhouse gas methane into our atmosphere," she wrote. "We need to develop truly clean energy, not dirty water created by fracking."[22]

Industry documents show no such thing. Statistics from environmental regulators show no such thing. Nowhere, anywhere, does any credible scientific evidence exist that fracking has made a single drinking water source "dirty." On the contrary, a review of tens of thousands of wells, in state after state, and by the most rigorous federal environmental regulators, has turned up a complete blank on any fracking-related drinking-water contamination.

It is no overstatement to say that fracking has proven 100 per cent safe for drinking water in the United States – making fracking probably one of the few resource-based industries on earth that can actually boast such a statistic. How galling it is, then, that so much of the anti-fracking movement relies on spreading the opposite of that fact – on spreading an outright lie.

Chapter Five

CRITICISMS OF FRACKING: USING TOO MUCH WATER

Fracking's enemies rely a lot on scaring people about their water supplies. And it's one of the most effective deceits they can deploy. Water is among the most fundamental necessities of life. We drink it, cook with it, wash with it, and so do our children. The fear of contaminated water is primal – animals are naturally wary of stagnant, dirty water – as is the fear of running out of water. And that's the other fear that anti-frackers have tried whipping up. Not only will your water be rendered unusable by fracking, goes their terror-propaganda, but soon you'll run out of it, too!

Fracking is "already straining U.S. water supplies," they warn us.[1] Towns are "drying up" from all those fracked wells.[2] We are "intentionally wasting . . . this precious life source."[3]

Of course, it's true that fracking requires a lot of water. It's the water that drillers need to force down deep into the shale. It is, actually, the vast majority of the fracking fluid that is pumped deep into the well. But that same fracking fluid that anti-frackers will tell you is a "lethal cocktail" of toxic chemicals is actually 99 per cent water, which they will then turn around and point out is a "precious life source."

Each fracked well uses millions of gallons of water. That's the statistic you'll hear over and over again from fracking's haters. And it's true. An average fracked well will use between two million and five million gallons of water.[4] That amount might seem unfathomable. But it's a drop in the bucket compared to the amount of water we use for everything in life, on a daily basis. In Fort Worth, Texas, alone, a city of just over 750,000 people, residents sprinkle an estimated 15 million gallons' worth of water onto their lawns every single day.[5]

Those lawns don't stay watered, mind you: they require constant, repeated watering, to stay green and healthy. But that's not the case with fracked wells. Once a well is fracked, it stays fracked for years. Those few million gallons that are pumped into that shale rock are used one time, before the gas is released, and the well starts producing. Once it's producing gas, the water-pumping stage is done: the trucks pack up and leave, the well is capped and attached to a pipeline, and it releases gas for years – and if it's fracked again, even decades. So that amount of water is effectively all the water that's needed for the entire lifetime of the well.

That means that energy from shale uses less water than many other forms of oil and gas extraction. Look how it compares to ethanol, which has been promoted by renewable-energy backers as a more environmentally sound renewable form of energy. It take 750 times more water to develop the same amount of ethanol fuel energy as shale-gas energy.[6] But, then, agriculture – which is where corn-based ethanol comes from – requires gargantuan volumes of water. In the United States, farms drink up about 128 billion gallons of water each and every day.[7] Industrial and mining water use in the United States amounts to about 3.5 billion gallons every day.[8] That's the kind of context we need to think about when we talk about the 2 million gallons that is used to frack a gas well.

Farming and business activities in New York, Pennsylvania, and West Virginia use vastly more water than all the water used in all the Marcellus Shale deposit. The U.S. Department of Energy estimates that at the peak of Marcellus drilling activity in New York, Pennsylvania, and West Virginia, there will be "maximum water" use of 650 million barrels of water per year.[9] Colossal? That amount of water accounts for less than 1 per cent of all the water used in the three states for other purposes, every year. This incredibly profitable and booming industry, this super-abundant supply of clean, affordable energy, will require all of one one-hundredth of the water being consumed for farming, business, homes, and utilities. Hardly the horrific, drought-causing menace that fracking-haters work so hard to portray.

Now, the average person doesn't realize how much water
we use every day for so many things. Why would they? And
so they don't know how to put two million gallons of water
– to frack a well – in context. Two million sounds big,
anyway. And the anti-fracking zealots use that to exploit our
lack of familiarity with water issues. They take advantage of
it. How many people really know how much water they use
in a day? We might have some vague sense of how much
we drink and use when we shower. But we consume huge
amounts of water doing so many other things we don't even
think about. It takes 634 gallons of water to make a single
hamburger[10] – to grow the cattle to make the beef. Multiply
that by the millions of hamburgers eaten across the country
every day, and suddenly the water needed to frack a well or
two, or five hundred, seems minor by comparison. How
many simple cotton T-shirts do you have? Each of those
simple cotton tees took 713 gallons of water to make – to
grow the cotton, and then to wash it in the manufacturing
process. That's billions of gallons of water each year used to
keep our fellow citizens adequately T-shirted. And that's
nothing compared to blue jeans, which can take more than
twice as much water – 1,800 gallons – to make a single pair.

If you had a mind to, you could launch a powerful pub-
lic-awareness campaign against jeans and T-shirts that
could frighten people far more about water usage than
any anti-fracking attack. Remember that the two million
gallons go into a fracked well only once, and then the well
is good to go, providing critical energy for heating and
cooking and jobs for the rest of its lifespan. It's likely those

wells will keep producing long after we replace our latest T-shirt or newest jeans.

That doesn't mean that we should be panicked about how much water jeans or T-shirts use either – it means that we have (and use) far more water than we imagine. The anti-fracking activists try to frighten us with a number like two million gallons. Even in a relatively "dry" state like Texas, that volume of water wouldn't keep the state's garden hoses flowing for even half an hour.

In fact, the anti-fracking militants have taken such advantage of our unfamiliarity with water usage that they try to fool us into believing the very opposite of the truth. Because fracking actually saves water. Because the water is used only once, at the start, fracking is one of the most efficient uses of water for fossil-fuel extraction that exists. A 2010 Harvard University study concluded, after analyzing water usage in different fossil-fuel contexts, that shale gas replacing fossil fuels was a "beneficial trend" as far as water use goes, because, despite the large upfront investment of water, over the life of the well, for the amount of energy produced, "shale gas . . . has lower water consumption than other fossil fuels."[11] Marcellus wells have even been shown to waste less water than conventional natural gas, the non-fracked kind: about 35 per cent less.[12]

And that's the context that matters most. This can't be a competition between fracked shale gas and some magical, zero-impact energy source that we haven't discovered yet. It's about how fracking compares to the actual alternatives – to other fossil fuels or to ethanol (renewable energy technologies, like solar, aren't a reasonable alternative, since you can't run vehicles on solar energy, but in any case, the making of photovoltaics uses significant amounts of water, too: a typical manufacturing facility might use more than 70 million gallons of water a year).[13] And in every case, quite contrary to the reckless misinformation being spread by the anti-frackers, shale-gas wells use less water to produce more energy than any of the fuel alternatives.

And that already impressive water efficiency is improving all the time. Fracking has been around for decades, but today's version – multi-stage horizontal fracking, with powerful 3-D seismic computers, improved mechanical precision, and so on – is a revolutionary technology. With it, the fracking industry has essentially been reborn in just the last few years. As with any technology like this, more efficiencies are worked out to save money while reducing impact (the two typically go together), and the amount of water required to frack a well is virtually guaranteed to be lower in five more years than it is today. Improvements are already unfolding at a remarkable pace. Drillers have reduced the amount of chemicals added to the fracking water, for example, and they're working to find ways to capture the "flowback water" – the water that gets pushed back up to the surface in the fracking process – and re-use it.[14] "The majority of companies are working toward reusing 100 percent of their flowback water," Dave Yoxtheimer, hydrogeologist with Penn State University's Marcellus Center for Outreach and Research, told the *Pittsburgh Post-Gazette* in 2011. "Environmentally it makes sense, and economically it makes more sense." The flowback is actually preferable to fresh water anyway, he said, because with all the dissolved metals and minerals it brings back up, it's "heavier" and more effective when you use it again for fracking shale. Range Resources, one of the first companies to frack in the Marcellus Shale, is already recycling close to 100 per cent of its waste water.[15]

But the future of fracking may not even involve the use of water at all: a company called Gasfrac,[16] as the name implies, uses a form of non-toxic propane instead of water to crack open the shale. The technology isn't in some deep research lab – it's been used well over a thousand times and its proponents claim it works better than water fracking. It may be that the industry's ceaseless drive for technological improvements does far more to protect the world's water than any loud-mouthed environmental activist.

Chapter Six

CRITICISMS OF FRACKING: SECRET CHEMICALS

The "hydro" in hydraulic fracturing means water – the water that's forced deep down the well bore, to crack open the shale rock miles beneath the surface. That water – the "chemical" H_2O – makes up close to 99 per cent of the "chemicals" used in fracking. So anti-fracking activists have naturally focused on the highly diluted additives – the couple of percent of fracking fluid – in their claims that the companies are poisoning the land and water.

After all, if they weren't poisoning the earth – if there was nothing to hide in their chemical cocktail – then why were the fracking companies hiding it?

The answer the fracking companies gave at first was an honest one, even if it didn't sound good: their fracking fluid recipes were as much a corporate secret as the formula for Coca-Cola or the Caramilk secret. If Halliburton or Range

83

Resources published to the world their exact selection and proportion of chemicals used in each well, that would give away proprietary trade secrets to their competitors.

In fact, fracking companies haven't even been allowed to keep their chemical mix a secret, at least not from first responders. Since the 1980s, the EPA's Community Right to Know Act required detailed lists of any toxic chemicals used in industry to be divulged to authorities.[1] The only special exemption is for bona fide trade secrets; those trade secrets still have to be disclosed to the local government, just not to the public.

But skeptics of big business are not always mollified by the fact that big government is acting as the watchdog. Conspiracy theories ran rampant, stoked by claims, such as that made by Josh Fox, the director of *Gasland*, that the whole thing was rigged by former vice-president Dick Cheney, the former CEO of Halliburton.

What would Coca-Cola do if a well-financed, professionally run campaign was launched against its secret recipe, claiming it was toxic? The company would surely dig in its heels at first. But after a while, if local city councils started voting to ban Coca-Cola, if conspiracy theories started circulating rumours that nefarious ingredients like "carbonic acid" were being included (a fancier phrase for carbonated water), Coke might simply decide to publish its recipe and hope for the best.

And that's exactly what the fracking industry decided to do.

Individual fracking companies, like Halliburton and Range Resources, have set up searchable websites with the precise recipe of fracking fluid for every well they drill. The composition is slightly different in each one – and, unlike the conspiracy theories, they seldom have more than a dozen chemicals in each mix – but it's all laid out for anyone to review, whether it's neighbours, scientists, journalists, or competitors.

Halliburton's own website[2] includes an interactive map, showing the different fracking fluids used, state by state. The website lists all the fracking fluid additives, no matter how minuscule the amount, and publishes the "Material Safety Data Sheet" – the sort of thing that would be filed with the EPA. But in addition to all that, the company does a bit of PR of its own. It describes what the chemical really does – showing that the reader has probably come into contact with it before, even in their own household.

Ammonium acetate sounds pretty dangerous. It's in Halliburton's fracking fluid recipe. But it's also found in other recipes. According to the disclosure page, its common uses include "Soups and Broths." As in, you eat it. Hydrochloric acid sounds pretty bad, too. But it's found in "table olives." It's a bit tougher to get worked up about something that's in your fridge, too.

Halliburton lists dozens of different formulations. But Range Resources[3] goes even further – you can simply type in the exact well you're interested in, and find out, down to the litre, what went into it.

"We are hopeful that our voluntary disclosure will help dispel the misconceptions that have persisted and allow Range and others to deliver on the potential of this extraordinary resource base," the company wrote when setting up its disclosure in July of 2010 – months before Josh Fox's movie *Gasland* was released into theatres claiming fracking fluid was a poisonous secret.[4]

A typical Range Resources disclosure form on its website lists every operational detail imaginable, and a few more.[5] The exact location of the well and the exact amount of water used is reported, down to the gallon – as well as where the water came from, and how much was recycled. Every company or contractor involved is listed, with their address and phone number.

The percentage breakdown of the chemical recipe is published to absurd precision: to eight decimal points. Water and sand typically make up 99 per cent of it, but inquiring minds can also find out, for example, that isopropyl alcohol – commonly called rubbing alcohol, and used as a disinfectant on cotton balls – made up 0.000049013 per cent of a particular fracking cocktail.[6] Conspiracy theorists won't be scared or scandalized. They'll be bored to death – or maybe they'll even call up one of the company phone numbers and ask for a job.

Many fracking companies have banded together in a single clearing-house-style website, called FracFocus.org where over 52,000 wells are disclosed, along with state-by-state fracking regulations and other useful information. Canada's two main fracking provinces, Alberta and British Columbia, have an identical website, at FracFocus.ca.

These websites aren't sexy; Hollywood celebrities haven't endorsed them and probably haven't spent time reading through them. They're not scandalous; in fact, they defuse the scandal through a level of voluntary disclosure that's probably unique among any competitive industry in North America.

In fact, the chief question they raise is this: How did the fracking industry come to use so many kitchen ingredients to drill for oil and gas?

Chapter Seven

CRITICISMS OF FRACKING: SEISMIC ACTIVITY

F racking is short for "fracturing" – as in, breaking rocks open to release oil or gas. It happens miles beneath the surface of the earth, but for many that doesn't make it sound any less dangerous – just more mysterious, or even unnatural.

As David Suzuki, environmentalist, scientist, broadcaster, and now anti-fracking celebrity, told a national TV audience in Australia,[1] "Fracking is one of the dumbest technologies there is. We have no idea what is under the ground. You know, because we are an air-breathing landlubber, living on the top skin of the planet, we think out of sight there is nothing down there."

That's an odd thing for a scientist to say – almost medieval in its superstition. Does Suzuki think there are elves or dragons down there? Maybe that we'd release the fires of hell itself?

Of course, mankind has been digging into the earth since before recorded history. Breaking into the ground for valuable things is what separated the Stone Age – when humanity basically picked up rocks that were lying on the surface of the earth – from the Bronze Age, when people dug into mines for copper and other useful metals. Archaeologists have dated the oldest copper mine to seven thousand years ago, in Serbia.[2]

Early mines weren't very deep, but that changed in the Industrial Revolution, when steam power made digging and drilling easier. These days, the deepest mines in which men work are literally miles beneath the surface. The TauTona gold mine in South Africa is four kilometres[3] (almost 2.5 miles) under the earth, a one-hour series of elevator rides. That's much deeper than most fracked wells.

But unmanned holes have been drilled much deeper than that; the Soviet Union started its Kola Superdeep Borehold project in 1970,[4] reaching more than 12 kilometres underground, or 7.6 miles. And an even deeper well was drilled in 2012 by ExxonMobil, to produce oil and gas.[5]

It seems absurd to have to retell the story of digging holes and drilling wells, but when anti-fracking activists speak as if drilling a hole and breaking rocks will release some sort of earth-demons, it's necessary to remind people that civilization and mining have gone together almost as far back as civilization and agriculture.

But still: fracturing sounds rough; it's breaking things; we think of "fracturing a leg" or another bone. Maybe Suzuki's superstitious warning doesn't hold up, but what about science? Does creating cracks deep in the earth risk any natural phenomena that might hurt people? Could fracking in California cause the famous San Andreas fault to slip?

Does fracking cause earthquakes?

Yes, it does – if only because the definition of an earthquake is any measurable movement in the earth. According to the U.S. Geological Survey (USGS), the American government agency that monitors quakes, "several million" earthquakes occur every year, with 1.5 million of those measuring 2.0 or more on the Richter Scale.[6] On any given business day, the USGS detects fifty blasts from coal mines large enough to set off seismographs.

But earthquakes below 3.0 on the Richter Scale are too gentle to be felt at the surface; that's why only scientists with specialized sensors can detect them.

Which is good, given the number of industrial processes that regularly cause tremors. Like "The Geysers," a cluster of geothermal power plants north of San Francisco. They pump water into a super-hot underground reservoir (heated by molten magma lower down). The steam that is created spins the turbines that produce about 1,000 MW of power.[7] But the act of pumping 20 million gallons of water a day[8] deep into a boiling underground cauldron creates a nearly constant series of quakes. The U.S. Department of Energy has a website with a real-time 3-D map of those quakes[9] charting their exact location and size. In a typical day, more than forty are measured, but most are too small for people to feel.

There's probably no other patch of land on earth as closely monitored as The Geysers – even though a dangerous quake there is highly unlikely. But other industrial earthquakes have been disastrous. In 1967, an earthquake[10] measured at 6.3 on the Richter Scale in Koyna, India, killed 177 people and left 50,000 homeless. According to the USGS it was a case of "reservoir-induced seismicity – earthquakes caused by the filling or changes in water level of large reservoirs." The underground faults just couldn't handle the stress of millions of tons of water backed up by a dam.

And the epicentre of the 2008 earthquake in Sichuan, China – 7.9 on the Richter Scale, killing 80,000 people – is just 550 yards away from the 515-foot-high Zipingpu[11] megadam, built in 2004, that holds back 315 million tons of water. Fan Xiao, the chief engineer of the Sichuan Geology and Mineral Bureau, said it was "very likely" that the dam caused the quake.

So it can happen: human activity can cause earthquakes. Especially massive new dams built over fault lines. But fracking doesn't use explosives, like many coal mines do; it's not a permanent, sustained pressure, like dam reservoirs. It's a burst of water pressure (and, in new variations of the technology, the use of pressurized air) that cracks open the shale rock once, typically in fractures smaller than a millimetre – as mentioned earlier, that's why sand is added to the fracking fluid, to prop open the tiny fissures.

And then the fracking is done – and the natural gas flows for years, even decades.

Fracking has never caused an earthquake of any size

– the largest ever detected could barely be felt by humans. But that hasn't stopped journalistic sensationalism, especially in places where fracking isn't well known, so tabloid journalism still works on the subject. "Fracking DID cause 109 earthquakes in Ohio, confirm scientists as opposition to controversial process grows," screamed the headline in London's *Daily Mail*.[12] That headline is probably all that most readers would see. But if they were to read more, they'd find that, sentence after sentence, the headline seemed less and less accurate.

"New research officially confirmed that 'fracking' caused the set of nearly a dozen mysterious earthquakes in Ohio in 2011," starts the story, moving from 109 to "nearly a dozen." The next sentence then mentions they were "small quakes" – too small to be felt, except through sensitive equipment.

But a reader would have to get halfway through the story to realize that the micro-quakes weren't caused by fracking at all. "Earthquakes were triggered by fluid injection shortly after the injection initiated — less than two weeks,' Columbia University seismologist Wo-Young Kim told LiveScience on Wednesday."

That still sounds like fracking – the fluid injection. But a look at Dr. Kim's original remarks shows he wasn't ascribing the quakes to fracking, but to the disposal of the waste water that is produced after the fracking process is complete.[13] It's true that the disposal of waste water in deep underground caverns can cause seismic activity – scientists have known about that since the earthquakes under the Rocky Mountain Arsenal, near Denver, Colorado, in the 1960s.

That arsenal was a testing facility for all manner of U.S. military projects, and it buried its waste water more than two miles under the surface of the earth – as much as 20 million litres a month.[14] After years of lubrication, the tectonic plates started to slide – minor quakes between 3 and 4 on the Richter Scale, which were large enough to be felt by people on the surface and to deeply concern residents of nearby Denver.

Those quakes had nothing to do with the military testing at the arsenal, just as the much smaller quakes in Ohio had nothing to do with fracking. It was how much water was disposed of after the fact, and how quickly. The lessons from the Colorado scare were that waste-water disposal – from any source – needs to be done at a rate that takes into account the seismic structure of the rock it's being injected into.

That's a long way away from the screaming headlines of the *Daily Mail*. But how many Britons buying their morning tabloid at the newsstand would take the time to track down Dr. Kim's study to read for themselves?

Everything is a seismic event, from a woodpecker pecking to a basketball bouncing to an underground nuclear test. It's a form of trickery to report on micro-quakes with the sort of screaming headlines reserved for the real tragedies. But when there's so little else for fracking critics to go on, what do you expect?

Chapter Eight

WHO ARE THE ANTI-FRACKING ACTIVISTS?

M any of today's most ardent anti-fracking activists used to love the stuff. What changed?

Back in 2009, Bill McKibben, the anti-fossil–fuels crusader, published his book, *Eaarth: Making a Life on a Tough New Planet.* He daydreamed about various fantasy scenarios to get America off oil (like plants that "eat nuclear power waste" and the greatest unicorn of all, low-cost solar panels). But amidst these unrealistic ideas, a real, workable idea managed to slip in: McKibben noticed the uptick in natural gas production in the United States – the start of the fracking revolution. Here's what McKibben wrote[1] back then: "The last year has seen new discoveries of natural gas in the United States that could help wean us off dirtier coal" and "at least in the United States, we've found some new supplies of natural gas, which is a good 'bridge fuel' between dirty coal and clean sun – you can retrofit your coal-fired plant to burn the stuff."[2]

McKibben didn't just talk about the environmental benefits of natural gas – he took to the streets about it, too. In 2009, he was one of the organizers at a civil disobedience rally[3] in Washington, D.C., where eco-protestors blocked the doors to the Capital Power Plant, the coal-fired heating and cooling plant that has served most congressional buildings for over a century.

McKibben and his crowd didn't call for nuclear-waste-eating plants – those are apparently still being worked on by Hollywood screenwriters and aren't quite ready for real life. They called for the plant's power supply to be switched from coal to natural gas – a proposal actually made by congressional lawmakers before McKibben's rally. And four and a half years later – lightning speed, in Washington – the Capital Power Plant finally got its permits to allow it to switch to natural gas.[4]

That Capitol Hill protest was just a few years ago; McKibben was so passionate about natural gas back then, he wrote an essay[5] about the protest called "Why I'll Get Arrested to Stop the Burning of Coal." Moving to natural gas was so important, McKibben was willing to break the law.

That was then. But now McKibben has turned on natural gas. The 2014 McKibben would probably want to put the 2009 McKibben in jail. His once-busy campaign website, www.CapitolClimateAction.org, has been scrubbed from the Internet altogether, although traces of it are visible on archived Internet caches.[6]

The main campaign slogans he had back then – to get rid of coal – are still valid today. Natural gas has a lower carbon footprint than coal does, and fewer other kinds of pollution, like smog-creating particle pollution. As the United States has switched en masse from coal-fired power plants to gas-fired power plants, America's greenhouse gas emissions have plunged and air quality has improved. Isn't this exactly what McKibben had hoped for? Why won't he take a victory lap – America is indeed embracing natural gas as a transition or "bridge" fuel, until his fantasy-fuel scientists can invent dilithium crystals, or whatever else he had in mind.

So what's the problem?

Well, that is the problem – that natural gas really was a solution. Economically, politically, environmentally – it all made sense, enough for America to actually embrace it. Not out of a passionate hatred for coal, which seemed to motivate McKibben. But out of self-interest – gas is cheaper, mainly. And America now produces a lot of it. The fact that it's cleaner is a bonus. That's called a victory, isn't it?

But that's the thing about victories. They put professional activists out of business. Being an activist for women's suffrage was a busy job in the 1910s. Then in 1920, the United States passed the 19th Amendment granting women the right to vote. It was huge news, great news. But, from a narrow perspective, all those activists had to find something else to do for a living. Same thing for ending the military draft.

Only in the professional environmental movement would good news be treated as bad news – would activists not take yes for an answer.

Perhaps they never really meant to propose natural gas as a solution in the first place – when McKibben wrote his book, natural gas was very expensive, dominated by Russia, Iran, and Qatar, and it looked like the next massive import project to the United States, like OPEC oil.

Maybe it's personal – a sense of postpartum depression, that coal is being phased out and greenhouse gas emissions are plummeting in the United States, and so McKibben doesn't feel as morally indispensable as he once did. Perhaps it's a funding issue: McKibben's big donors, including anti-fossil–fuel foundations like the Rockefeller Brothers Fund, are ideologically committed to fighting against every last carbon molecule, and there's just too much campaign money to pass up.

Or perhaps it's something philosophically deeper than that: the very idea that we could have a plentiful, affordable, high-energy lifestyle that doesn't damage the planet is somehow a rejection of the underlying moral message of radical environmentalism, that our industrial civilization is immoral. The simple slogan "reduce, reuse, recycle" isn't just a three-point to-do list, it's an ethical statement: that humankind should just do less, use less, consume less, maybe even live less.

It often makes sense to reduce, reuse, or recycle – no point in being wasteful, and finding more efficient ways to do things saves time and money. But environmentalists like McKibben said we needed to live smaller in order to save the world, that our large living was precisely what was causing the planet to heat up. Natural gas has done

everything that McKibben hoped it would do in 2009 – it
has reduced America's GHGs to levels so low they haven't
been seen since the early 1990s. Isn't that good news? We
can call off the rationing!

Unless the rationing was the whole point – and the
global warming was just the excuse. Maybe some environmentalists weren't so much about saving the world, as they
were about condemning and correcting the behaviour of
their fellow citizens.

Chapter Nine

COAL MINING VS. FRACKING

There isn't an industrial process anywhere that comes without some impact on the environment or without some level of risk. Any commercial enterprise, whether fruit farms or server farms, has some effect on the environment – disturbing land, using wood and minerals for construction, consuming electricity, creating waste, or any number of other alterations of nature. And some of the most hazardous industries in our modern economies are actually old-fashioned, basic kinds of work: fishing, logging, and farming.

Even environmentalists understand that impacts are unavoidable. That's why they rely on things like carbon "offsets" – paying money for low-carbon efforts happening somewhere else, to compensate for the CO_2 they cannot avoid producing themselves. Al Gore uses as much electricity every year in his 10,000-square-foot Tennessee mansion as seventeen average American households use, but he says it's okay because he compensates by paying poorer people not to use energy (it just so happens he buys those credits from a carbon-offset-sales company he also partly owns).[1] Carbon offsets are the modern-day equivalent of the Church selling indulgences to rich people who sin; they can't stop, so they use their money to excuse their "sinful" behaviour.

In *Gasland Part II*, the sequel to his original anti-fracking propaganda film, Josh Fox focuses on a man named Steven Lipsky who shows off to the camera how he can turn his garden hose into a flame-thrower, just by putting a lighter to the fumes that he claims are emanating from the water inside. But Lipsky is a fraud. He was publicly exposed long before Fox released his sequel. In a lawsuit hearing involving Lipsky, a Texas court found that he had intentionally attached the hose to a gas vent, not a water line, "to provide local and national news media a deceptive video, calculated to alarm the public. . . ."[2]

But just compare the fastidious – and dishonest – hunt for a single speck of a problem with fracking to the effects we have grown to accept as entirely "normal" when it comes to other energy sources, such as coal.

Coal power is still the largest source of electricity in the United States, and it really does pollute. Coal-fired power plants emit about twenty different kinds of potentially toxic chemicals, including arsenic, lead, mercury, nickel, and radium. They send up sulphur and mercury emissions. They create smog and acid rain.

Coal comes with a death toll, and not an insignificant one. We don't even need to look at pollution, which involves an indirect link between bad air and deadly illness. We just need to look at the actual number of people who are dying in the coal industry, where the death of coal miners is, unfortunately, considered one of the costs of doing business. Even China is willing to admit that somewhere between three and four thousand coal miners die every single year (which is probably an understatement, knowing the Communist dictatorship's notorious reluctance to reveal unflattering statistics).[3] In the United States, roughly 100,000 people have died in coal-mining accidents in the last century.[4] Sometimes they make the news, when it's a spectacular enough disaster, such as West Virginia's Upper Big Branch Mine explosion in 2010, which killed 29 miners. But even the years that have record low deaths – 2009 was the lowest with 18 coal-mine deaths, though the following years saw that number multiply again – still chalk up many more coal-related fatalities than the dramatically safer practice of shale-gas fracking.[5] About twenty people still die every year.

This isn't a polemic against coal; for much of the world, especially in developing countries like China, coal-fired power plants are the best hope millions of people have to be lifted out of poverty. But, like all industries, it has its risks – and after centuries of coal use, we in the industrialized West have come to understand this fact of life. It's absurd to hold a newer source of power – affordable, clean fracked gas – to a new fantasy standard of perfection.

Opponents who want to stop fracking for shale gas act as if the alternative is consequence-free. Shale-gas fracking is a choice – a choice over other kinds of energy extraction, whether that's coal, which actually causes real pollution and real death, or imported energy from countries that use their export revenue to pollute and kill somewhere else. Shale gas isn't perfect – no fuel is. There is no perfect energy source that comes without some impact on the environment and some risk. But shale-gas fracking is arguably the best choice on offer, at least until the anti-fracking lobby presents us with the faultless fantasy fuel that we've all been waiting so long for. Until then, shale gas is better. Not perfect, but better.

Chapter Ten

WHAT IS FRACKING REALLY LIKE IN AMERICA?

Wh, with its financial sector, on a rebound from the
the United States? Which Canadian province
has the lowest unemployment rate?

In the era of Facebook, could it be California – home of
Apple, Google, Twitter, and hundreds of other high-tech
giants? Or maybe Boston, with its constellation of Ivy
League universities and the hot jobs that go with them?
New York, with its financial sector, on a rebound from the
Great Recession?

Not even close. Even Facebook's IPO, which created a
thousand new millionaires,[1] hasn't made a dent in
California's staggeringly high unemployment rate of 8.9
per cent,[2] higher than the national average. The tech boom
has been wonderful for top executives or those lucky
employees who got in on the ground floor of a company

whose app hit the big time. But that's a sliver of a fraction of the general population.

Massachusetts' unemployment rate is a touch better, at 7.2 per cent,[3] pretty much the U.S. average. But it's heading up, too.

How can Boston – a capital city, a port city, a banking city, home of Harvard and MIT – have such a mediocre economy? How can New York – the world's banking centre, media centre, transportation centre – have an unemployment rate even higher, at 7.6 per cent?

And how does North Dakota – a state not known for its universities or financial industry or ports or really anything particularly money-ish – have an unemployment rate of 3 per cent, one that's actually still falling? This is the state whose sexiest tourist event is the Norsk Høstfest – billed as North America's largest Scandinavian festival. The official state dance is the square dance. They have an official state grass: the Western Wheatgrass. Is that their secret?

No; fracking is. North Dakota is in the middle of the Bakken geological formation, which produces both oil and gas through fracking. And unlike the banking jobs of Manhattan, or the dot-com jobs of Silicon Valley, it's not just the CEOs and early shareholders who are making out like bandits – it's anyone who drives a truck, knows how to weld, or even runs a motel or a restaurant.

Same thing in Canada; it's not coincidence that the Canadian province with the lowest unemployment rate is Saskatchewan – the one that borders North Dakota. Its unemployment rate is just 3.8 per cent[4] too. Both North

Dakota and Saskatchewan have small populations – fewer than two million people between them. But that's changing quickly, as young men and women looking for jobs stream in.

It's not a coincidence, it's a pattern. Even in other states, unemployment rates and economic growth, on a county-by-county basis, closely match the amount of fracking activity.

A study[5] prepared by Diana Furchtgott-Roth, a former chief economist of the U.S. Department of Labor, found that over a four-year period, Pennsylvania counties with more than 200 fracked wells experienced per-capita income growth of a whopping 19 per cent – despite the national recession. Pennsylvania counties with only moderate fracking activity – between 20 and 200 wells – saw their income rise by 14 per cent. Counties with fewer than 20 fracked wells saw 12 per cent income growth. And in those counties without fracking, incomes were up just 8 per cent between 2007 and 2011.

That's just income. Employment was directly correlated with fracking, too: in counties with more than 200 wells, job numbers grew at 7 per cent per year. In counties with no drilling, jobs numbers shrank by 3 per cent.

It's ironic: in the 2013 election in the province of Nova Scotia, the central issue was how to bring back jobs and stop young people from leaving for greener pastures. Outside the main city of Halifax-Dartmouth, unemployment was above 10 per cent, hitting a whopping 15.2 per cent in Cape Breton.[6] All three parties said bringing the jobs back was their priority. But all three parties agreed that the province should maintain its moratorium on fracking – just to be safe. They never seem to make the connection. But the young workers from Cape Breton packing up for a better life in Saskatchewan do.

ROYALTIES TO LAND OWNERS

The United States boasts the richest consumer economy in the world and has some of the most plentiful arable farmland of any country. But a year after British and Irish rock stars formed the Band Aid supergroup to sing "Do They Know It's Christmas?" to raise money for famine-starved Ethiopia, Willie Nelson, John Mellencamp, and Neil Young started their own charity concert – for American farmers.

Ethiopia hasn't had a famine since 1985 – and the Marxist regime that exacerbated the mass starvation with calamitous agricultural policies and military spending is long gone (though the country still often finds itself barely able to feed its people). But it's been almost thirty years since Farm Aid started and it's still going strong. Performers have held three dozen concerts now, in Illinois, Texas, Nebraska, Indiana, Louisiana, Kentucky, South Carolina, Virginia, Pennsylvania, Ohio, New Jersey, Washington, Missouri, and New York, raising tens of millions of dollars in handouts for American farmers. That's on top of the subsidies north of $10 billion that Washington doles out to agriculture every single year.[7] The heartland farmer, once a symbol of American self-sufficiency and capability, has become a perpetual national charity case.

How that happened is a complicated story, but it is more accurate to say that it's a certain kind of American producer that's struggling: the small-scale, family-owned farm, which actually sees very little if any of those subsidies. Larger agri-business consolidators are the ones lobbying for, and getting, the most government largesse, and they're the ones that, for various reasons – including distorted subsidies – the small farmers have trouble competing with.[8] No one wants to see small American farms failing, and yet, they are. The average American farm family relies heavily on other sources of non-farm income just to stay afloat.

Nowadays there's a rapidly increasing chance that that other non-farm income is royalties from shale development. Take the Georgetti family and their dairy farm in southwest Pennsylvania. Before the shale-gas boom hit, their dairy farm was on the financial brink. "We used to have to put stuff on credit cards. It was basically living from paycheck to paycheck," farmer Shawn Georgetti told an Associated Press reporter. With thirty-year-old equipment, he was falling further behind in productivity and he had no way to upgrade without taking on significant debt.

Then a company called Range Resources came and made a deal with the Georgetti family. Range would drill a small natural gas well on the farm and would pay the family a chunk of the revenue – the minimum royalty rate in Pennsylvania is 12.5 per cent.[9] Now the Georgettis can upgrade their machinery and they don't have to mortgage the family farm to do it. "It's a lot more fun to farm," Shawn said. No kidding.

The royalty windfall from shale-gas fracking is pulling back untold numbers of small family farms from the brink of insolvency. Royalties are called that because they began as the rent a producer of some land-based commodity – say, silver or gold – would have to pay to the king, who owned the land. That's still the case in a lot of places, such as Canada, where most of the mineral rights belong to the government, whose head of state is Queen Elizabeth. Anyone mining or drilling for oil or gas in Canada pays a portion of the money they make to the province or, less often, the federal government. But in the United States, the royalty system is largely private. Had Jed Clampett been living in Alberta when he shot a hole in the ground and saw that bubbling crude come up, he never would have made it to Beverly Hills (producers in Canada typically pay a relatively minor lease payment to use private land for their drill pads, but the farmers who live on top of the land are entitled to no portion of the income from the minerals underneath). In the United States, though, there are Jed Clampett stories happening all across shale country – troubled farmers becoming instant millionaires thanks to the unleashing of the shale-gas resources they've been sitting on top of all these years.

Allen and Debbie Francis have been scraping out a barebones living on their land along the Kansas–Oklahoma border for twenty years. One horizontal well on that land means as much as $500,000 every month from shale oil and gas revenues, Debbie told ABC News.[10] Six million dollars a year. And what will they do then? Keep farming, Allen Francis says. "I wouldn't quit. That's all I know is farming." Though he did buy a new $174,000 tractor. Not everyone's making quite that much. More common are stories of families making tens of thousands of dollars in royalties a month, still a huge difference to those farms. And a lot of them are using their money as Allen did – to improve their farms. Business is booming at the John Deere dealership in Anthony, Kansas, as farmers plunk down hundreds of thousands of dollars for the machinery they could only dream of before.

"Many of the farmers I deal with every day are customers that we've dealt with for twenty years. Some of them have wanted to have new tractors for a long time and haven't been able to afford it," says Pat Myers, general manager of the John Deere dealership. "It's nice to see these guys able to do this after they've struggled for so long."

It's an incredible turn of luck for these farmers. But it's also an incredible turn of luck for the institution that is the American family farm. Farmers have been an integral pillar of the American identity, but more importantly than symbolism, they actually provide an important service: they make food. And for so many of them, it's what they want to keep doing. They find themselves sitting on a jackpot, thanks to shale fracking, and what do they do? They turn around and buy equipment that will allow them to farm even better than before. Fracking is, as one economic development official in a farm-reliant part of Wisconsin put it, "the biggest and best thing that's happened in our lifetime."[11] In Mountrail County, North Dakota, the epicentre of a shale-fracking boom, the average per-capita income doubled in five years, rocketing it into one of the hundred richest counties in the nation.[12]

Precise statistics on how many millionaires are being made aren't available – these are private transactions, so we have to estimate. But in June 2013, the Pittsburgh-based fiscal think-tank, the Allegheny Institute for Public Policy, estimated that royalties in the Marcellus Shale region had exploded, from $10.9 million in 2010 to $731 million just two years later.[13]

That's just one region, in one country, where shale gas is being tapped. Most countries are unlike the United States, with its private royalty system. In Britain and Canada, and many European nations, minerals are publicly owned and gas producers pay royalties to the government. But in those cases, private producers pay billions of dollars in taxes that end up funding school districts, hospitals, and other social services. The results are less concentrated, more dispersed. But in all cases, the result is that underground gas that was no good to anyone before can suddenly help power economies with affordable, clean power, while distributing revenues that can make a whole lot of people a whole lot better off. From Pennsylvania to Poland – that is the moral result of fracking.

LOWERING CO_2 EMISSIONS

In the five years from 2007 to 2012, U.S. greenhouse gas emissions – the culprit in the theory of man-made global warming – fell by a whopping 12 per cent, back to 1995 levels, according to the U.S. Department of Energy. As Dr. Jeffrey Frankel, a Harvard professor who served on President Bill Clinton's Council of Economic Advisers, asks, "How can this be? The United States did not ratify the Kyoto Protocol to cut emissions of greenhouse gases below 1997 levels by 2012, as Europe did."[14]

In fact, the U.S. government's Energy Information Agency says that during the spring of 2012, U.S. emissions were the lowest in twenty years – since 1992.[15]

But back in 1992, the United States only had 255 million[16] people. How could the emissions today – with 316 million people, with their cars and homes and businesses – be less?

The recession of 2008 is a small part of the answer – America's economy shrank. But that recession ended in 2009. Since then, America's economy has grown, but its carbon dioxide emissions have kept shrinking (while Europe's increased).

"Something else is going on," writes Frankel. "The primary explanation, in a word, is 'fracking.' In fourteen words: the use of horizontal drilling and hydraulic fracturing to recover deposits of shale gas."

Burning natural gas in power plants releases only half the carbon dioxide that burning coal does. And as fracking has brought the price of gas down low, the use of gas in power plants since 2007 has risen a whopping 37 per cent. Coal is down a quarter.

It wasn't a foreign treaty or a carbon tax that did that. It was the free market, substituting cheap, plentiful, clean natural gas as a power source.

In fact, the opposite is happening in Europe. They don't have the economic incentive to switch to natural gas power plants, because they don't have fracking. And the political drive to shut down European nuclear power, as a knee-jerk emotional response to the tsunami that hit the Fukushima reactor in Japan, has forced German power companies to turn away from zero-emission nuclear reactors, back to coal. Ironically, the Kyoto countries are growing their emissions, while the United States – which signed, but never ratified the treaty – leads the way in reductions.

Chapter Eleven

SHALE GAS AROUND THE WORLD

S hale is one of the most common types of rock formations on earth. It's a sedimentary rock, made by layers of mud compacted together over time. It's easily broken apart, in layers. And between those layers, or in pores, natural gas can be trapped. That's what fracking does: in a moment of pressurized fracturing, it makes cracks in the rock that allow gas to escape, over years or even decades.

Fracking is an American technology, and the U.S. government's Department of Energy (DOE) has studied the geology of fracking intensively, not just in the United States but around the world. In 2011 and again in 2013, the department published an authoritative world map showing the location of shale-gas deposits around the world, in forty-one different countries.[1]

In many of those forty-one countries, detailed geological information just doesn't exist, but the DOE used the best available information they could about 137 different geological formations. Including the United States, the DOE estimated that shale-gas reserves around the world total 7,300 trillion cubic feet of gas (and 345 billion barrels of shale oil, too). In the two years alone between editions of that study, the DOE's world estimates for shale gas grew by more than 10 per cent.

The DOE's map of shale gas is a striking contrast to traditional energy maps, where a handful of OPEC countries dominated the world supply. Other than plain old sand, shale is one of the most common rock formations, and the DOE has found major gas resources in every continent. From northern Alaska, down through Mexico, Brazil, and to the southern tip of Argentina; from South Africa to the Sahara Desert; from Britain and France to Russia and China; and Australia and Pacific islands, too, shale gas is everywhere. There are gaps in the DOE map – central Africa, central Asia, and the far north of Siberia. But that's likely a function of scant research data, as opposed to unlucky geology.

The top ten list of lucky countries is almost completely opposite the traditional OPEC map. China, one of the world's largest importers of energy today, has the largest shale-gas reserves in the world (1,115 trillion cubic feet), with Argentina in the second place (802 tcf). Algeria, in third, is the only OPEC country on the top ten shale-gas list, followed by liberal democracies: the United States (665 tcf),

Canada (573 tcf), Mexico, Australia, and South Africa. Russia is in ninth, with Brazil in tenth. Other estimates published by the DOE suggest that the United States actually has more shale gas than any other country, but whether it's in first or fourth place, America finds itself in a startlingly different position than its desperate dependence on imported crude oil, much of it from OPEC dictatorships.

In the United States, shale-gas recovery is most developed in states like Texas and in southeastern states like Arkansas and Louisiana. The Bakken Formation in North Dakota has given that state the lowest unemployment rate in the country, a microscopic 3.0 per cent.[2] The Appalachian Basin in the U.S. northeast has rich shale-gas plays like the Marcellus Shale, in West Virginia, Pennsylvania, and New York. Shale gas is in Michigan, Colorado, Illinois, and Oklahoma. There's even shale oil under Los Angeles.

Shale gas is plentiful, and it's found in liberal democracies just as often as in authoritarian regimes. It's not an exaggeration to call shale gas "freedom gas." And even countries that aren't in the top ranks of shale gas often have game-changing amounts – like Poland, whose 148 tcf is enough to liberate that country from the predations of Russia's state-controlled gas exporter, Gazprom.

POLAND

During the Second World War's pivotal Yalta Conference, U.S. president Franklin Roosevelt made an offhand remark to Winston Churchill, wartime prime minister of the U.K., that Poland "has been a source of trouble for over 500 years."³ That may have been a slight exaggeration, at least as far as the United States was concerned, but the history of Poland is certainly a troubled one, even if the Polish people were far more often on the receiving end of trouble, rather than the ones causing it.

It was especially fitting that Roosevelt would make such a remark on the occasion of the Yalta meeting with Churchill and Russian dictator Josef Stalin. It was February 1945, and the war was in its final throes in Europe. Within three months, Hitler would be dead, and German forces would surrender Berlin. As much as anything, the Yalta conference would determine the future of a Europe freed from Nazism. But for the Poles, it would represent only a trade of one vicious tyrant for another: Soviet forces had liberated Poland from the Nazis only to seize it for Stalin. For the next forty-four years, the Poles would be captives of the Soviet Union – owned and operated by, and for the benefit of, Moscow.

Poland had suffered brutally under the Nazis during the Second World War, more than any other nation conquered by the Nazis. Hitler considered Poland to rightly belong to Germany, and he was determined to wipe out the Polish people's very identity – the plan was called "dePolonization" and "Germanification." Before the Nazi invasion, which ultimately triggered a world war, Hitler told his generals that his goal was not to capture any specific region or strategic asset, but "the destruction of the enemy."⁴ To Hitler, the Poles were "racially inferior," a subspecies, like the Jews.

Polish citizens were deported to remote areas or concentration camps to make room for the resettlement of Germans. They were banned from school. Members of the Polish intellectual class – writers, poets, artists, academics – were tormented and marked for death. Poland lost more Jews to the Nazi death machine than any other country. Of the 3.3 million Jews who lived in Poland on the eve of the Nazi invasion, fully 3 million were murdered, nearly 10 per cent of the country's population and more than three times as many Jews as were murdered in Ukraine, the country with the second-highest Holocaust death toll.

But as cruel as Hitler's occupation was for the Poles, falling back into Moscow's clutches would hardly have seemed a "liberation," let alone a relief. As Roosevelt so indelicately pointed out, Poland's troubles date back centuries, much of it caused by Russia. In the seventeenth century, Peter the Great colonized Poland, and in the eighteenth century it was carved up by Catherine the Great, with part going to Russia, part to Prussia, and part to Austria. In the nineteenth century, it was Napoleon taking parts of Poland for France. Poland saw true independence for only a few years, between the First and Second World Wars, but the twentieth century was surely Poland's bloodiest of all. And after the Second World War, Stalin's totalitarian repression picked up where Hitler left off in his attempts to obliterate the Poles' national identity and enslave them to a foreign master. The Soviets persecuted the Church and confiscated Polish farms and businesses, while Poland's real history was once again censored and made illegal, in order to dePolonize the country – this time as part of an effort to assimilate it into the Soviet Union.

Poland comes honestly by its yearning for independence and freedom. This is the country that produced some of history's greatest anti-Communist heroes: Pope John Paul II, Lech Walesa, and the pro-freedom Solidarity movement. Having spent nearly its entire existence being invaded, subjugated, carved up, dominated, and brutalized by outsiders, Poland is, in many ways, still working to undo the damage and chart its own independent national course. As Donald Tusk, the Polish prime minister and once a leading activist in Poland's anti-Soviet Solidarity movement, explained to *Time* magazine a few years ago, the disastrous effects of Russia's Communist rule have yet to be completely cleared away, and the country cannot waste any time, he said, in rebuilding (certainly history teaches that any weakness makes Poland only that much more vulnerable). "We have no oil and gas," Tusk said. "We don't have high tech. Our centers of development, are far, far behind others. We will never be an extraordinary tourist attraction. Poland is quite a mediocre country in some regards. The only natural resource that we have, and with which we can compete, is freedom."[5]

That was in 2008. But in the handful of years that have passed since, Tusk's assessment of Poland's resources have been proven to be just a wee bit off the mark. A 2012 analysis from Ernst & Young reported that Poland had the "largest reserves of shale gas in Europe" at an estimated 792.81 trillion cubic feet, and at a minimum 187.17 of it technically recoverable.[6] Eventually, the U.S. Energy

Information Administration pegged Poland's reserves at 148 trillion cubic feet.[7]

Poland, under Tusk, has embraced shale-gas exploration. And it's no surprise why: Poland had been almost totally reliant, like so many Eastern European countries, on Gazprom's Russian gas monopoly. In 2010, Poland was the number one importer of Russian gas, with 80 per cent of the country's demand relying on Gazprom's pipeline[8] – which, given Poland's long history spent being kicked around by Moscow, you can be sure hasn't sat well with the Poles. Nor has Gazprom proved to be particularly good for the Polish economy that Tusk and previous governments have been trying to resuscitate after the ravaging disease of Communism. Like so many former Soviet vassal states utterly dependent on Gazprom, Poland had been rewarded with exorbitant gas prices.

Poland pays staggeringly high prices for its energy, like $5.50 a litre for gasoline, or about $20 a gallon. And while it's possible for poor people to take public transit instead of driving a car, everyone has to pay to heat their homes. That's where the exorbitant price of natural gas really hurts Poles: $12.75 for a thousand cubic feet of the stuff.

For comparison, the average price of natural gas at the wellhead in the United States in 2012[9] was $2.66 per thousand cubic feet,[10] and during the spring the spot price fell as low as $1.84. That's not a typo: Poland was paying eight times more for its Russian Gazprom gas than American utilities were paying at Louisiana's Henry Hub.[11] That's the most important natural gas trader in America, supplying to thirteen different pipelines and setting the price benchmark for the rest of the continent.

That new Polish price of $450 is after their national gas utility, PGNiG, managed to negotiate a 20 per cent price reduction[12] from Russia's state-controlled Gazprom. Even the new, discounted price is so unbearably high, PGNiG has to sell gas to customers at less than what it pays Russia for it. The Polish government has to subsidize the price of gas, or its citizens will freeze.

It's a massive transfer of wealth from Poland to Russia, and it's all because Russia's Gazprom has a near-monopoly on gas imports to the country.

Except it's worse than that. In February of 2013, the Russian newspaper *Izvestia* compiled a map of all of Gazprom's customers in Europe showing how much gas the Russian giant sold to each country, and at what price.[13] What was immediately clear was that there was no geographic or economic rationale for the fact that Gazprom sold gas to Poland at $14.87 per thousand cubic feet (*Izvestia* was using the price of Gazprom's existing contract, not the newly negotiated· slight discount of $12.75), but sold the same gas to Germany – farther away from Russia, literally on the other side of Poland – for just $10.48. The gas that went through Poland to get to Germany was cheaper than the gas that stopped in Poland.

It's the same metal tube, called the Yamal-Europe pipeline.[14] Same natural gas molecules being pumped through it. But Vladimir Putin charged customers in Poland 38 per cent more for the same stuff. Why?

An analysis done by *Forbes* magazine shows that there is no economic rhyme or reason for it – the graph of countries and prices is just a scatter. But that's because *Forbes* is a financial magazine. Gazprom's pricing is political – not surprising, given that its majority shareholder is the Kremlin.

The countries that are charged the most – coloured in bright red on the *Izvestia* map – just happen to be former Soviet satellites, countries that used to be under Warsaw Pact military and political domination by Moscow. Poland had it the worst on the price list, but the Czech Republic and Bulgaria are also at the astonishing price of $14.16 per thousand cubic feet. Serbia, Romania, and Slovakia were also on the list of the highest-charged Gazprom customers.

On the low end were the U.K. – ironically, the furthest destination to which Gazprom had to ship – and the Netherlands and Germany, though still being massively gouged by Putin.

Why did Gazprom charge the Brits $8.87 per thousand cubic feet, but Poles $14.87? Bad negotiating skills by the Poles? Hardly. You can't negotiate in a hostage-taking situation, when you yourself are the hostage. Gazprom's token price reduction was little more an attempt to fend off a European Union inquiry into price-fixing.

The Poles are as sensitive as anyone to the abusive practices of Gazprom and Moscow's habit of using gas for cynical geopolitical gain. When Gazprom surprised Tusk in early 2013 in announcing it would be building new pipelines across Poland (a political move in itself, since Gazprom was looking for ways to bypass Ukrainian territory, part of Gazprom's ongoing discrimination against Ukraine), Tusk insisted that he would not allow Poland to be bullied any longer by the Russian gas monolith. "Poland won't participate in these political contexts," Tusk said. "For us, gas isn't a tool to conduct politics and we very much want, in agreement with European Union laws, to keep gas issues free of politics."[15]

Before the shale-gas revolution, Poland had tried to diversify its energy mix. That ended up "spectacularly unsuccessfully," as *The Economist* bluntly put it.[16] "As a result they have been forced to accept gas import prices higher than those paid by their richer western neighbours." Once fracking opened up the possibilities of not just an energy-independent Poland, but the possibility that Poland might become an exporter of gas, displacing Gazprom supplies from other European import markets, the government moved quickly. The first exploration well was drilled in 2010, and the country now has more wells in action than any other country on the continent.

Exploration companies have since found shale-gas recovery in Poland to be a bigger challenge than in North Dakota: the deposits are deeper, and infrastructure to the more remote parts of the country, such as roads, has yet to catch up to the modern age.[17, 18]

Those poor roads and inferior infrastructure are, in large part, the vestiges of Russia's inept and corrupt Soviet rule over Poland. But the Poles have faced more severe tests than hard-to-reach shale deposits. To the explorers trying to make a go of things in Poland, shale gas may seem a challenge. To the Polish people, it can only seem like an inevitable final step to freedom.

UKRAINE

The history of Ukraine is a history of a people ceaselessly trying to free themselves from Russia's grip. Even the name "The Ukraine" literally means "the frontier" of Russia. To Russia's rulers – from the tsars, to Vladimir Putin and his stewards at Gazprom – Ukraine is a hinterland. A satellite. A colony. A possession.

Few other countries in history can make the kind of tragic claim to subjugation and brutality under another country's iron fist. More than two decades after Ukraine declared independence from Soviet rule, Russia continues to treat Ukrainians as wayward vassals, subjects of Moscow's regional domination. There's even a word for this: "Russification." For centuries, Russia had attempted to stamp out, to ethnically cleanse, all signs of Ukrainian culture and language. Russian leaders wanted to take their belief, their fantasy, that Ukraine did not even exist as a separate nation, and turn it into reality. They maintained that there was no such thing as a Ukrainian people or a Ukrainian tongue. The tsars banned the printing of Ukrainian-language publications. The Soviets murdered almost every member of the Ukrainian cultural elite: writers, artists, intellectuals, and priests.

That's what makes Ukraine's ability to tap its shale deposits so momentously historic. The Energy Information Administration now calculates that Ukraine has 128 trillion cubic feet of recoverable shale gas – just a bit behind Poland's and France's resources. With Ukrainians currently importing 1.6 tcf of gas every year from Russia, the country has the ability to displace all its Russian imports for as much as eighty years.[19]

It is hard to overstate the kind of turning point fracking represents for Ukraine. Because Ukraine doesn't just pay Russia for gas, the way a normal customer pays a supplier for gas. There is also the painfully expensive cost that comes when you're buying from a monopoly whose objective isn't to serve you as a customer, but to ensure its customer remains servile. Every so often, Gazprom announces, rather out of the blue, that Ukraine suddenly owes it billions more dollars in unpaid extra charges. In January 2006, in the middle of an Eastern European winter, Gazprom demanded more money from Ukraine. Putin claimed Ukraine was "stealing" gas from his gas monopoly. And to punish Ukraine, he told Gazprom to cut off all gas supplies to the entire country.

If you live in a cold, Western country, it's likely illegal for the gas company to just cut off your heat in the middle of winter even if you refuse to pay your bill; often there are rules against it in colder months, or at least the gas supplier has to give you plenty of notice, so you aren't hit with a sudden and surprise freeze one cold January night. But Gazprom? It cut off gas to an entire nation of 45 million people. Just like that. Because it can. Because, in Russia's eyes, Moscow remains in charge of Ukraine, and if it means cutting off the entire population from heat, cooking, even power (about 20 per cent of Ukraine's electricity comes from natural gas), so be it. In 2007, Gazprom, claiming it was owed $1.3 billion in unpaid charges, threatened to do it again.[20]

You could probably devote an entire book to the conflict between Ukraine and Russia over gas contracts. That's not this book, but the seriousness of the problem – *The Economist* in 2009 said the ongoing tensions between Ukraine and Russia over gas had "grown into the biggest energy emergency the European Union has seen in years"[21] – underscores just how monumental something like energy freedom is to people. And there is probably nowhere where it means more than in Ukraine.

For many countries, energy freedom comes with huge economic benefits. For countries like the United States, it comes with that, but also with geopolitical benefits – the end of enriching regimes that are hostile to American interests. But in Ukraine, energy freedom means finally bringing an end to a literally centuries-old regional oppression. For Ukraine, the benefits aren't just economic and geopolitical – they're psychic.

Russia's desire to dominate Ukraine is in its very blood. It sees Ukraine as its rightful territory. When Ukraine declared independence in 1991, after the collapse of the

Soviet Union, that did not stop. In 2004, when Ukrainians took to the streets to protest government corruption and election rigging by the Kremlin-friendly president, Viktor Yanukovych, Russia didn't just intervene in lending political support and staff to Yanukovych and spreading praise for him across Russian media, while portraying the protestors — numbering in the hundreds of thousands — as misguided or under the influence of Western agents. Russia is widely thought to have been behind the poisoning of the opposition leader, Viktor Yushchenko.

He was poisoned at dinner with Volodymyr Satsyuk, deputy head of the police agency that replaced the Soviet KGB in Ukraine. At that dinner, Yushchenko ingested massive amounts of a toxic dioxin and became gravely ill. His face became horribly and permanently disfigured. Only three labs had been producing the particular formula of dioxin that medical examiners found in Yushchenko's system. Two of them produced their samples to investigators. The only lab that refused was based in Russia. After fleeing to Russia, Satsyuk was given Russian citizenship and protection from extradition to Ukraine.[22]

The possibility that Russia was involved in such ruthless intrigue could hardly surprise anyone (two years later, former Russian spy Alexander Litvinenko, who had fled Moscow for Britain, died from radioactive poisoning also linked to Russian agents).

But it's important to remember just how brutally Russia can, and has, dealt with Ukraine in the past. Remember that Ukraine was on the receiving end of one of the worst genocides in modern history, perpetuated by Russia. The Holodomor – the "hunger extermination" – was Josef Stalin's starvation of the Ukrainian people from 1932 to 1933.

That was when Ukrainians, already struggling with drought and crop disease, were forced into collectivized farms to fulfil Stalin's communist vision. Stalin ordered all the "kulaks" – the successful farmers – liquidated, claiming they were capitalist oppressors (in reality, they just knew how to farm properly), and while farm outputs dropped, he confiscated Ukrainian crops to supply Russia with food. Many scholars agree the famine was a way for Stalin to subdue Ukrainians chafing under Soviet domination. He certainly dismissed any attempts by those close to him, including his own wife, to bring relief or aid to the Ukrainians dying en masse.

The exact number of Ukrainians who starved or were killed in the Holodomor may be incalculable, though most scholars today put it somewhere between three and five million people. Fourteen countries have recognized the Holodomor genocide, including Canada, the Vatican, and Australia. Five more have recognized it as a criminal act by the Stalinist regime. And in 2011 the Democratic National Committee in the United States adopted a resolution recognizing the genocide and supporting an American monument in remembrance to it.

Ukraine, once victimized by Russia's collective-agriculture oppression, is today being victimized all over again by Russia's gas extortion. If the Russian government had truly undergone a thorough cleansing of its Soviet past, the situation might be at least somewhat less torturous for Ukraine. But remember that Vladimir Putin was an agent of the Soviet regime. He was an active and enthusiastic participant in Soviet repression as a KGB agent. And even today, Putin is working to whitewash Stalin's horrors, with "a comprehensive program to ideologically reeducate society" by trying to "minimize" Stalinist crimes, reports Russia's independent Levada Center polling group. Under Putin, support for Stalin as a "positive" leader is strong, with only a third believing his influence was negative, a 2012 Levada poll found. That's the reverse of what Russians thought before Putin began working to rehabilitate Stalin's image. In 2010, the U.K.'s *Daily Mail* discovered that Russia's new textbook for teachers maintained that Stalin committed his murders "entirely rationally" as "the guardian of a system."[23]

Imagine if Israel were held as an economic captive by a German regime that was trying to whitewash the history of the Second World War. That gives you a sense of how humiliating and demoralizing it is for Ukraine to remain in the grips of the Putin regime. Gazprom is not committing another Holodomor. But there's something deeply immoral about threatening to starve Ukrainians of heat and cooking fuel in the middle of January. That's the alternative to fracking.

GERMANY

After a 2011 earthquake and tsunami crippled the nuclear power station at Fukushima, Japan, one of the first countries to announce that it would quickly abandon nuclear as a power source was Germany.

That was a politically bold move. It was also dangerous. In Fukushima there had been a coincidence of disaster and circumstance that could never happen in Germany. Japan sits on a famously volatile fault zone, where earthquake risk is highest. A triple fault line, actually – where the Eurasian plate, the Pacific plate, and the Philippine plate rub against one another, sometimes with violent results. Germany doesn't sit in an earthquake zone. It has had all of two quakes in the last fifty years. They were barely felt, and one of them was caused by mining activity. Japan has had thirty-one. Many of them were disastrous. Thousands have died in earthquakes and tsunamis well before Fukushima. Just as importantly, the nuclear meltdown at Fukushima was as much a result of the tsunami that followed – which swamped the generators responsible for back-up nuclear cooling – as it was due to the earthquake that disabled the Fukushima Daiichi power station in the first place. Germany does not get tsunamis.

But even in the compound disaster that hit Fukushima, a basic fact remains: no one died. Not from the reactor anyway – the tsunami killed people. But not the nuclear plant.

Still, Germany's powerful green lobby took advantage of the Japanese tsunami to stoke public fears against nuclear power, and the government panicked. Never mind that nothing like this could ever happen in Germany. Never mind that in Japan, even after that incredible series of events led to a meltdown, not a single death has yet been attributed to radiation leaking, there has been no rise in cancer rates since the accident (the area had been largely evacuated by the time radiation had leaked).

But the government of Angela Merkel had three upcoming state-level elections to worry about, and the environmental cause risked draining votes from her party. So seven nuclear plants were closed immediately. The remaining ten were to be phased out in the next few years. Before the announcement, nuclear power was providing a quarter of all German electricity. Germany had to find a way to make that up. Immediately.[24]

The trouble is, Germany is simply not very self-sufficient when it comes to that other incredibly useful power source, natural gas. Germany is the biggest energy consumer in Europe, excluding Russia, and the seventh-largest in the world. And while it produced 459.1 billion cubic feet of its own natural gas in 2010, it consumed 3,425 billion cubic feet – more than seven times as much, making it critically dependent on imported gas, mostly from Russia.[25]

Without nuclear, and heavily reliant on Gazprom to keep its powerful economy running, Germany has been searching desperately for alternatives. It's thrown billions of dollars at solar and wind power, hoping that the sun will shine and the wind will blow just consistently enough to keep its $3.5-trillion economy from blacking out. In 2011 and again in 2012, Germany narrowly escaped large-scale power blackouts. The sun and the wind, it seems, are not cooperating with Angela Merkel's plan to "transition" away from fossil fuels. But Germans may not be particularly cooperative either, lately. Additional charges consumers are being forced to pay to subsidize the renewable industry have increased by nearly 50 per cent. German electricity prices are now among the highest in Europe. The worst isn't even over yet: German environment minister Peter Altmaier warned in May 2013 that the cost of Germany's "Energiewende" program (literally "energy transition") could reach 1 trillion euros by 2039. That's more than 12,000 euros for every man, woman, and child currently in Germany, or more than US$15,000.[26]

Already Germany's energy-intensive industry sectors, such as chemicals, are paying four times as much to buy gas as U.S. industry pays. That's good for Germany's competitors in the United States, but it's not good for many others. Germany is the only truly healthy economy in continental Europe; that country is pretty much shouldering the bulk of the costs of the EU crisis. If you're interested in a healthy global economy, then you had better hope that Germany stays in good economic shape.[27]

As Germany phases out nuclear and even coal power and replaces them with unreliable, costly, and insufficient wind and solar power, most analysts see no way for Germany to stay in good shape without a great deal more natural gas power. And Germany has it, if it only cares to tap it. Its extractable shale gas reserves are estimated at approximately 17 trillion cubic feet. That doesn't mean Germany is on the road to eternal self-sufficiency, but it does mean it can dramatically slash its reliance on Gazprom imports.

So what's the problem? Well, it's that same green lobby that pushed Germany into this precarious energy-high-wire act in the first place. Not satisfied with demanding the end to nuclear and coal, Germany's environmental groups, like France's, are demanding an all-out ban on fracking. They've already succeeded in getting the opposition Green and left parties to propose bills outlawing fracking – though, luckily, those were defeated by the coalition led by Angela Merkel's Christian Democratic party. The Bundesrat upper house, representing the states, passed a resolution in early 2013 demanding that Merkel's government tighten rules for fracking. Right now there is an effective, albeit unwritten, moratorium on fracking as the government seems paralyzed with uncertainty on how to proceed.

That de facto moratorium on fracking looks set to continue for years; in fall 2013 elections, Merkel's party received its strongest showing since the merger of East and West Germany in 1990. But, as is usually the case in Germany, parties need to cobble together a coalition to govern. Merkel's traditional coalition partner, the pro-fracking Free Democrats, had a disastrous election, falling out of government and even being ejected from Parliament, for failing to receive 5 per cent of the vote.

The pro-frackers are out; Merkel's new political partners, the centre-left Social Democrats, have agreed to put shale gas on ice. It looks like Germany's dependence on Russian gas will continue for years to come.

FRANCE

France imports 98 per cent of its natural gas needs, or 4.7 billion cubic feet per day. A quarter of that comes from Russia.

But France's thirst for energy has driven it into even rougher hands than Vladimir Putin's. France was one of the largest importers of energy from Libya, under its late dictator Muammar Gaddafi. It was Western Europe that kept Gaddafi in power by buying energy from him, and that lucrative business financed his brutal dictatorship for decades.

And the thing about doing so much business with a corrupt dictator is that it just might corrupt you, too. According to allegations made before a French judge, France wasn't just importing Libyan oil and gas, but Libyan values: Gaddafi, claims Middle East fixer Ziad Takieddine, paid 50 million euros to finance Nicolas Sarkozy's 2006 election campaign. Sarkozy denies the charge.[28]

But there's no need to point to secret deals to demonstrate the problem with France having to import so much of its energy. The public facts of it are bad enough: at best, being dependent on other countries for energy is an economic drain, as billions of dollars a year leave the country. And at worst, it risks the security of supply of that energy – especially when so many of the suppliers are Arab dictatorships, and the so-called Arab Spring is still roiling.

In January of 2013, Al Qaida–linked terrorists attacked a major Algerian natural gas facility near the Libyan border. The hostage-taking lasted four days, and when it was over, thirty-seven workers at the plant were dead, including six Brits.[29]

Security analysts claimed that the attack on the gas facilities was a reaction to France's military action against Islamic extremists in Mali, another African country. Whatever the rationale, the attacks on European energy sources seem to be the new normal: only a few weeks later, two pipeline guards were killed in another terrorist attack on Algerian pipelines.[30]

But France is lucky; it sits on enormous, proven resources of shale gas, in large geological basins around Paris and the southeast of the country: 727 trillion cubic feet are in place, of which 137 tcf is estimated to be recoverable at current prices and with today's technology.[31]

A million of anything is hard to fathom, and a trillion is a million millions. But put it this way: if France is sitting on 180 tcf of gas, that's a little bit more than 100 years of the country's total natural gas needs. And the same French shale that holds promise for gas has oil in it, too – 117 billion barrels of it in place, with up to 4.7 billion barrels recoverable with today's technology and prices,[32] in the Paris Basin alone.

But fracking in France is at a standstill. Perhaps it's because the technology is publicly associated with America – France is the country that loves to hate (or at least pretend to hate) American symbols like Walt Disney or McDonald's. But here's the thing: after noisily condemning Euro Disney for years, the French embraced it, so much so that it's now the top tourist attraction in all of Europe.[33]

But fracking is definitely in the "hate it" category for now. José Bové, a Green Party politician and an anti-McDonald's activist, has whipped up opposition to fracking through a potent combination of environmental fear mongering and good old-fashioned retail politics.[34]

Bové achieved quite a feat: France's super-major oil company, Total S.A., had already received permits for fracking and was starting to assemble drilling rigs outside Paris. Other companies from the United States and Canada were doing the same, including Vermilion Energy, which had already fracked several wells. There were sixty-four fracking permit-holders in France when the government pulled the plug.

Even mighty Total couldn't fend off the Green surge: the promise of a domestic, secure supply of oil and gas just wasn't persuasive enough, even in an economy with unemployment over 10 per cent, the highest level in fifteen years.[35]

It didn't help much that France's political point man on fracking at the time, Eric Besson, sounded more like a Greenpeace activist than the industry minister his business card claimed he was. Fracking chemicals, Besson said, "have caused considerable damage in the U.S. and Canada." That would come as a surprise to Canadians and Americans, but imagine the impact such a confession would have on an ordinary Parisian who was uncertain about fracking: if the industry minister himself admitted it was dangerous, it surely was as bad as Bové said, if not worse.

Not that Sarkozy himself, the president at the time, was any better. When he finally announced the moratorium on

French fracking in 2011, in the run-up to the 2012 election that he lost by just 3 per cent, Sarkozy decided to outflank the anti-fracking opponents on the left. Shale gas could come "at the price of fragmenting the soil that would massacre the almost spiritual scenery" of France's mountain landscape, he warned. Forget the fact that fracking happens thousands of feet underground – not on massacred mountains. France was against it, and Sarkozy wasn't going to die on that hill.

But although fracking is banned in France, that hasn't stopped Total S.A. from fracking – in America. The French super-major paid close to a billion dollars to buy 25 per cent of Chesapeake Energy's assets in Texas, and it's actively planning to frack from Canada to Poland to Algeria.[36]

UNITED KINGDOM

The United Kingdom was the seat of the first great Industrial Revolution, and that country has forever been associated with coal. But the government's decision in late 2012 to let fracking proceed could make clean-burning natural gas the economic mainstay in the U.K.

In fact, coal will likely not be the biggest loser if shale gas catches on: massive wind turbines will be. The U.K. embraced the promise of wind energy with a great passion, spending billions of dollars subsidizing giant, skyscraper-sized turbines that blighted the landscape across Old Blighty. But the promise has not been fulfilled; wind farms don't work when it's not windy. Even the most efficient turbines still need a conventional back-up source of power. And then there are the environmental and personal objections – the devastation caused to birds and bats, the noise and visual pollution, and claims of illness from vibrations. Across the U.K., dozens of grassroots groups[37] have formed to oppose wind farms; they are the exact opposite of the international, jet-setting environmentalists-of-fortune who make up the professional activists at groups like Greenpeace. These local objectors have been increasingly successful;[38] in 2005, just 29 per cent of wind factories were opposed by local planning boards. Five years later that had risen to 48 per cent. By the fall of 2012, the U.K.'s energy minister, John Hayes, was telling newspapers that "enough is enough," suggesting that wind turbines – at least those on land – were facing a moratorium, just as fracking was getting started.

For all of the criticisms of fracking, it doesn't come close to the intrusiveness of wind turbines – there is no constant humming, vibration, or visual ugliness; no daily harvest of unlucky birds and bats; no 400-foot-high metal monstrosity in what was once a cornfield.

And as to the energy part of the energy industry, natural gas doesn't need a back-up generator in case it doesn't burn. It *is* the back-up generator.

According to the blue-ribbon British Geological Survey,[39] the U.K. is sitting on a staggering 1,300 to 1,700 trillion cubic feet of shale gas – 200 times more than previously estimated. That would be enough to heat every home in the country for more than a thousand years.

Fracking may sound newfangled and exotic to some in the U.K., but it has been used in that country for decades; fracking, along with directional drilling, developed that country's oil fields in Dorset in 1979,[40] where two hundred wells have been drilled. It's the largest onshore oil field in all of Europe. British gas fracked a well in Lancashire back in 1996 that has been producing natural gas ever since.

As the prospect of a fracking boom loomed in the U.K. press – and as professional environmentalists-of-fortune started ramping up their anti-fracking campaigns – the Royal Society and the Royal Academy of Engineering released a report on the subject, called "Shale gas extraction in the UK: a review of hydraulic fracturing."

Now, anyone can issue a report – and generating scientific-sounding propaganda is a specialty of the anti-fracking lobby. But the Royal Society is different.

Founded in 1660, the Royal Society of London for Improving Natural Knowledge is the country's most respected scientific advisory body. Its motto, *Nullius in verba*, roughly translates to "take nobody's word for it." For 350 years, its elite scientists have promoted empiricism – that is, science through observation and experiment, not through politics.

The Royal Society's fracking report, issued in June 2012, was unambiguous: fracking was safe, provided that standard regulations would be in place. As their report noted, "The UK has 60 years' experience of regulating onshore and offshore oil and gas industries" already.

Fleet Street had been feasting on the more tabloid-friendly aspects of hydraulic fraturing – just think of all the naughty headline puns that could be written with the word "fracking" – but the one that caught their attention the most was the risk of earthquakes. In the spring of 2011, a year before the Royal Society's report, sensitive seismic monitors detected tiny earthquakes near a fracking project next to Blackpool, in the northwest of the country. The *Independent*'s headline was one of the more restrained: "Exclusive: Fracking company - we caused 50 tremors in Blackpool – but we're not going to stop."[41] The *Daily Mail* wasn't going to be out-head-lined:[42] "'Man-made' earthquake strikes Blackpool . . . and consequences could be severe for UK's gas drilling industry." And Rupert Murdoch's *The Sun* simply said: "Blackpool rocked."[43]

But did it really rock? What did the *nullius in verba* folks – the don't-take-anyone's-word-for-it folks – have to say?

It's true, the Royal Society wrote, that a fracking well near Blackpool likely caused the quakes. It was picked up by some of the hundred seismic monitoring stations run by the British Geological Survey. But the magnitude of the quakes – 1.5 and 2.3 on the Richter Scale – were so low that they were almost undetectable.

"Given average background noise conditions in mainland UK, a realistic detection limit of BGS' network is magnitude 1.5. For regions with more background noise, the detection limit may be closer to magnitude 2–2.5. Vibrations from a seismic event of magnitude 2.5 are broadly equivalent to the general traffic, industrial and other noise experienced daily," they wrote.

That's what the tabloids' rock-and-shock fracking quakes were like – one of them was barely detectable by scientific instruments, the other was like a car passing on the street. The idea that man-made activity would cause an earthquake may seem unsettling, but as the Royal Society noted, it's been happening in the U.K. for centuries: "The energy released during hydraulic fracturing is less than the energy released by the collapse of open voids in rock formations, as occurs during coal mining. The intensity of seismicity induced by hydraulic fracturing is likely to be smaller due to the greater depth at which shale gas is extracted compared to the shallower depth of coal mining."

Less energy released, at a much deeper depth than coal mining – the Royal Society says the chances of any real problems are negligible. There are literally hundreds of such micro-quakes a year in the U.K., and no one feels them.

Still, one of the mandates of the Royal Society is to provide scientific advice to the government, and it did, coming up with a checklist of sensible regulations, based on its own science, as well as the experience in U.S. fracking. They recommend some of the same regulations used by the "green" geothermal industry, which also causes micro-seismic mini-quakes. They all fall under the category of common sense – mapping out naturally occurring geological fault lines in advance; monitoring seismicity before, during, and after drilling; having traffic-light–style warning systems for when to shut off fracking. None of it's new – the industry is decades old. But for the world's oldest panel of experts to give their sign-off on the safety of fracking not only gave the government the confidence to proceed, it might even force tabloids to explain, ten paragraphs after a screaming headline, that a basketball game is higher on the Richter Scale than fracking is.

BULGARIA

In 1944, Soviet troops invaded Bulgaria and claimed it for the U.S.S.R. And Bulgarians have been trying to get away from Moscow's control ever since. But when Russia controls virtually every molecule of natural gas that you need to power your economy, ending that domination would seem unrealistic.

Since the collapse of the Soviet Union, Bulgaria has been turning, as much as it can, toward the West. Bulgarian leaders had begun looking to break away from the Warsaw Pact well before that, in the late eighties. In 2004, Bulgaria joined the United States and its allies in NATO. Shortly after that, it joined the European Union.

These are not merely symbolic nods toward the free West. Soviet-occupied Bulgaria was a founding signatory to the Warsaw Pact, along with other recent NATO allies, Poland and Romania. When those countries sign on with the very military alliance that stood opposite the Warsaw Pact, it is as powerful a repudiation of Russian regional influence as can be. And whatever its many flaws, the EU is a Western-facing organization. To a Soviet nostalgic like

Putin, Bulgaria's pro-American pivot is worse than a rebuke; it's a slap in the face.

It's a formal declaration that Bulgaria will never belong again to Russia or even be within Moscow's sphere of direct influence. It can't be. NATO, even today, stands consistently opposed to Russian interests. In Libya, while NATO intervention assisted and accelerated the overthrow of the Gaddafi regime, Putin refused to allow his UN representatives to vote in favour of the no-fly zone or to vote in favour of calling for Gaddafi to resign. He griped about NATO's involvement. "[W]hen the whole of so-called civilized society gangs up on one small country, destroying infrastructure that has been built over generations . . . I do not like it," he said. "What do they need to bomb palaces for? To drive out the mice?"[44] He accused the West of using the anti-Gaddafi rebellion as cover to raid Libya's oil and gas reserves. In Syria, Putin stood by another vicious Middle Eastern dictator, Bashar al-Assad, clashing with NATO's anti-Assad stance. And Russia consistently blocks Western efforts to press Iran into abandoning its nuclear ambitions.

As much as possible, Bulgaria is breaking free from Russia's clutches, to never again return. It keeps agreeable relations with Moscow, but strategically, economically, legislatively, and militarily, Bulgaria has made it clear it plans to chart its own pro-Western course.

Only, when it comes to energy supplies, Bulgaria remains trapped in Russia's grip. Bulgaria is shockingly dependent on imports for its natural gas – 94 per cent of the roughly

105.94 billion cubic feet of gas Bulgarians consume every year comes from Russia.

But Bulgaria has shale gas. How much shale gas? The U.S. Energy Information Administration (EIA) estimates it could have roughly 17 trillion cubic feet of shale gas, or about 480 billion cubic metres[45] to use metric (the Energy Delta Institute suggests Bulgaria may yet have double that much).[46] That means if the EIA estimates are even close to correct, Bulgaria has enough shale gas to be completely free of Russia's imports for the next 160 years.

Bulgaria almost certainly has enough gas to be energy self-sufficient for decades, or to dramatically slash imports (and its dependence on Russia) for even longer. Literally, for generations. But no one knows for sure just how much shale gas Bulgaria has – because Bulgaria won't let anyone find out.

Incredibly, Bulgaria has banned all fracking. And that means that even exploratory fracking wells, which would properly assess the amount of shale gas within Bulgaria, can't investigate. Bulgaria doesn't even know what its economic potential could be; it cannot calculate what the opportunity cost of banning fracking amounts to. Chevron had initially been granted a five-year permit in June 2011 to explore for gas in Bulgaria's 4,400-square-kilometre Novi Pazar shale field where initial estimates, based on similar geological formations, suggested there was the potential for between 300 billion and 1 trillion cubic metres of shale gas.[47] In January 2012, Bulgarian regulators cancelled Chevron's permit. Bulgaria had announced it was banning all fracking.

That might seem unconscionable, in a country where Gazprom had an iron-fisted monopoly on all natural gas. With that kind of power, Gazprom had simply dictated to the Bulgarian government how high gas rates would be. And Bulgaria had to take it; politicians in the capital, Sofia, had no bargaining power. For years, Bulgarians had been paying astronomically high prices for gas – up to four times the going market rate. And when Gazprom cut off gas to Ukraine in January 2009, demanding more money, Bulgarians, stuck at the end of the Ukrainian pipeline, froze for days, too. Why would they pass up a chance at finally casting off the last vestige of Russian domination?

Maybe because Russia arranged it that way.

That's what an editorial in Bulgaria's largest-circulation newspaper, *Dneven Trud*, insisted, immediately after the legislature approved the ban. In "Russian Lobby Against Shale Gas," editorial writer Ivan Sotirov was enraged at pro-Russian groups operating inside Bulgaria. It is why, he said, Bulgaria had yet to allow "a single serious strategic Western investor to set a foot in Bulgaria."[48]

That might sound a bit dramatic. But the editorial noted that the most powerful political opponents to fracking were those affiliated with the "Sixth Department" – Bulgaria's Communist-era secret service – and those connected to a governmental energy deal with Russia, including a new nuclear power station, an oil pipeline, and, of course, a Gazprom natural gas pipeline.

"The most shameful fact is the realization that after 22 years of democracy Bulgaria's policy continues to be dictated by oligarchic pro-Russian circles," Sotirov fumed. "[T]he National Assembly has banned Bulgaria from learning whether it has shale gas deposits – information which could have released us from the total energy dependence on Russia . . . which contradicts the Bulgarian national interest and protects our total energy dependence on Russia."

All of it, he wrote, had been done hastily and without any serious political pressure. There had been protests in Sofia, it's true. Three days before the Bulgarian National Assembly voted on the fracking bill, anti-fracking demonstrators took to the streets. Agence France-Presse reported that "about 1,000 youngsters marched along the streets of the capital Sofia," with smaller protests elsewhere in the country.[49] Sofia is a city of about 1.2 million people. So, about a tenth of a percent of the population took to the streets. And the government bowed to their piddly little parade? Banning not only the production but the very exploration for shale gas, so that Bulgarian citizens could make an informed decision about what kind of sacrifice a total prohibition would mean? And rather than heeding the lesson of decades of overbearing Russian control? Leaving themselves at the mercy of another shut-down pipeline, the next time Gazprom decides to cut off Ukraine.

"[A]ll this has been done without any serious motivation because the campaign against the shale gas prospecting has been based on cheap manipulations and lies," the *Trud* editorial went on. "This has been an attempt to disguise a political issue as a purely ecological matter."

How political? Two months prior to the sudden ban, when Chevron was still poking around to see the kind of potential Bulgaria might have for weaning itself off Gazprom imports, Gazprom announced it was buying up a chain of Bulgarian gas stations.[50] Before that, Gazprom revealed it was planning a series of new natural gas pipelines across Bulgaria, to increase access to and market share in Eastern and Central European markets.

And Bulgaria's most recent long-term natural gas contract with Gazprom was on the brink of expiring, when the legislature suddenly capitulated to a tiny protest group to ensure that Bulgarians would remain dependent on Russian gas imports for years to come.

Reporting on the Bulgarian economic and energy minister's decision to keep the ban in place until fracking could be proven "safe," Russia's state-owned international broadcaster, Voice of Russia, informed its audience that "according to scientists, [fracking] contaminates ground waters and may cause earthquakes in the earthquake endangered zones."[51]

Russia had an immense deal to lose if Bulgarians went ahead with shale-gas exploration – not just Gazprom's monopoly on Bulgarian energy, but potentially the proposed pipeline routes through Bulgaria, to Serbia, Romania, and Greece, that would eventually reach Hungary, Italy, Austria, the Czech Republic, and Germany.

In exploring for shale gas, Bulgarians, as the old Communist saying goes, had nothing to lose but their chains. And incredibly, their government took a pass.

It wasn't a total loss for Bulgaria. After such a near-miss with Bulgarian energy independence, Gazprom sat down to negotiate a new long-term contract with Sofia in a much more flexible mood. The Russians offered Bulgaria a 20 per cent discount on its previously inflated gas prices, with an option to renegotiate in five years.[52] Call it a quid pro quo – a modest reward for staying loyal to Uncle Vladimir, and throwing out those Yankees from Chevron.

Had the Bulgarian government played its cards better and opened its country to the possibility of shale-gas exploration, it wouldn't need to sign long-term contracts. It wouldn't need to ask for an option to renegotiate, or wait five years to do it. It could start supplying its own gas and dictate new import terms to Gazprom. For once in its history, Bulgaria could have been the one giving orders to Moscow.

LITHUANIA

As in Poland, Lithuania's modern history had known only brief moments of freedom from Russian rule, before the collapse of the Iron Curtain. Russia controlled Lithuania from 1795 until 1915. Twenty-five years later, Russia was back, this time in the form of invading Soviet troops. It was the Nazis that drove the Russians out again, in 1941 – the Lithuanians revolting against the retreating Red Army. Not that the Lithuanians were any more free under Hitler's rule. But soon enough, by 1944, the Soviets had returned, with no plans to leave again. By the time the Soviet system collapsed, more than 100,000 Lithuanians had been deported to exile in Siberia, the highest deportation of any country in the Soviet Union. Lithuanians represented just 1 per cent of the Soviet population and yet, in the final years of Stalin's rule, one-sixth of Soviet deportations were Lithuanian.[53]

Lithuanians resisted where they could; particularly in the countryside, attacks on the Soviets were fierce enough that Russians took to referring to the struggle as a "civil war" in Lithuania. But it was not until the dissolution of the Soviet Union that Lithuania could, for the first time since the late 1700s, confidently call itself free of Russian control. It was the first Soviet republic to declare itself independent. Still, today it remains reliant on Russia for its very economic survival. For its energy.

Lithuania is 100 per cent dependent on Russia for natural gas. And like so many other captive markets, and ones that Moscow still looks at as wayward colonies, it pays higher Gazprom prices than even harder-to-reach Western European countries do. On average, Lithuanians paid 15 per cent more for gas in 2012 than the European average, the second-highest rate on the continent, behind only Bulgaria.[54] In 2012, the Lithuanian government tried suing Gazprom to recover $2 billion it figures it has been over-charged for gas, going back to 2004.[55] It also pressed the EU to announce an anti-trust probe into Gazprom's discrimina-tory pricing practices in Central and Eastern Europe.

But lately, Gazprom has been showing a lot more flexi-bility with Lithuania. In September 2013, Gazprom chief Alexey Miller suddenly announced that his company was prepared to offer "a significant reduction in prices for

Lithuanian consumers."[56] The company even issued a press release, with the headline "Gazprom and Lithuania aim for mutually beneficial cooperation."[57] Isn't that friendly?

What had Moscow feeling so surprisingly generous? It might have had something to do with the fact that Lithuania is at long last on the verge of breaking Gazprom's longtime stranglehold on the country's energy supply.

President Dalia Grybauskaite had recently thrown her support behind a plan to build a floating liquefied gas terminal off the Baltic coast that could import gas from world markets, instead of being entirely dependent on Gazprom. The government estimates prices will be in the neighbourhood of 20 per cent lower than what Gazprom squeezes out of Lithuanians.

Grybauskaite is determined to free Lithuania once and for all from Russian dominance. But importing LNG is just part of her plans. She has also waved off the meddling of anti-fracking activists and ploughed ahead with plans to open up Lithuania to shale-gas fracking. The country is sitting atop what is estimated to be a substantial reserve of shale gas – potentially enough to supply all Lithuania's gas demands for a half-century or longer.[58] And the deposits are located twice as close to the surface as those in neighbouring Poland, making them that much more accessible and economical.[59]

The country has already started to work with Chevron to explore its own shale-gas potential. Lithuania, Grybauskaite said after meeting with Chevron officials, "must explore the depths of its land."

It won't happen overnight, but Russia no doubt is already realizing that its days of abusing Lithuanians are about to end. There isn't much it can do to stop the country from tapping its own shale-gas reserves, thank goodness, but it has begun trying to interfere with the LNG import plan. The Lithuanian president herself has suggested that Gazprom has been secretly working behind the scenes to meddle in the LNG project, in hopes of delaying it. But she won't be deterred. "To stop the project is not possible," she said, "but to intervene and to delay it partially, yes." Construction of the floating LNG terminal is scheduled to begin in 2014. But the rig already has a name. No surprise, it's Independence.[60]

CHINA

Affordable natural gas isn't just about improving individual people's lives, and our economies as a whole. It's about staying competitive with rising economic competitors. Like China.

Western countries have already watched huge chunks of their manufacturing industries decamp to China, which has been able to offer far cheaper labour costs. America alone lost five million manufacturing jobs between 2001 and 2011, more than half of them to China.[61] But that's actually starting to turn around. Labour costs in China are rising fast, as the middle class there grows, and their wage demands rise – the same effect that happened in the United States and so many other Western countries as they became richer. Manufacturers are already starting to return to the United States: the Boston Consulting Group estimates that by 2020, this "reshoring" effect could bring back millions of jobs to American shores.[62] The same effect could happen in Europe, too. One of the main reasons: cheaper energy, thanks almost entirely to the fracking boom. "The U.S. is steadily becoming one of the lowest-cost countries for manufacturing in the developed world," concludes the Boston Consulting Group's analysis.[63]

That's great for the United States. But what about those Western economies that have also lost countless jobs to Chinese outsourcing and can't bring them back with cheaper energy from shale-gas fracking? Places like Bulgaria and France, for instance?

Because make no mistake: China has no time for the bogus anti-fracking propaganda spouted by environmentalist mischief-makers. China has never worried about those industries that really do have serious environmental consequences. Its major cities are cesspools of pollution, so thick with smog that it's often dangerous even to leave your home. Fracking is as safe an energy-extraction technology as there is, though, so there is absolutely nothing standing in the way of China taking full advantage of it.

And the Chinese are. China, remember, is obsessive about becoming the most powerful economy in the world. And to that end, it has been intently driving toward securing its own energy supplies. It's moved swiftly and forcefully into Africa, working with any country it can find that has oil reserves, and helping them develop production, while securing exclusive trade deals.[64] China now gets about a third of its oil from African sources. Meantime, China's state-owned oil firms have been buying North American oil and gas producers wherever they're able. In 2011, China Petrochemical Corp. paid $2 billion – 70 per cent more than the company's average trading value in the preceding weeks – to acquire Canada's Daylight Energy Ltd., securing strategic Canadian oil and shale-gas reserves with the deal.[65] The following year China National Offshore Oil Co. paid $15.1 billion to buy Canada's Nexen Inc., including its stake in Canada's valuable oil sands.[66]

But China's most powerful energy advantage may be its shale-gas reserves. In 2011, the U.S. Energy Information Administration estimated China's recoverable shale-gas deposits at 1,115 trillion cubic feet, in just two basins – larger than all the reserves in the United States and Canada combined.[67] China began exploratory drilling in 2011 and has already signed deals with Royal Dutch Shell, Chevron ConocoPhillips, and Total S.A. to develop all that shale gas as much, and as quickly, as possible.[68]

Europeans are already paying a dear price for the costly fantasies of environmental utopians. Submitting to anti-fracking hysteria in the face of the cheap Chinese energy threat is only going to make matters that much more severe. In Germany, the government was pressured to move away from fossil fuels and nuclear by environmental lobbyists who promised that wind power and solar power would make for a "green economy" and magically extract no serious price from industrial productivity. Today, in the most energy-intensive sectors in Germany, businesses are having to pay four times as much for natural gas as do their American competitors, who are enjoying the benefits of the U.S. shale-gas revolution.[69] Germany's timidity about fracking can only make its industry increasingly uncompetitive once China's economy gets the added boost from cheap shale-gas energy.

Those jurisdictions that refuse to take advantage of fracking, the ones that pass up on the opportunity for secure, ample, and cheap energy inputs, won't be seeing any "reshoring" of industries lost to China. Whatever jobs they have already lost will have no reason to return, as China's cheap-labour advantage is replaced by another edge: a cheap-energy advantage. For those economies, the great export of jobs overseas certainly isn't going to reverse. It will, you can bet, only get worse. Places like Bulgaria, and France, or New York state, or Quebec, with their moratoriums on shale gas? They're going to see a lot more "Made in China" labels, and a lot fewer manufacturing jobs.

ISRAEL

Israel is the only liberal democracy in the Middle East, and it's surrounded by petro-dictatorships. For decades, that hasn't just been an economic handicap; it's been political too. Where other Western powers would normally have more in common with Israel, OPEC's Arab members have used oil as a diplomatic and economic weapon to tilt the regional balance of power in their favour.

That's not a conspiracy theory; it was the express rationale for the Arab oil shock of 1973, punishing the West, and especially America, for its support for Israel during that year's Yom Kippur War. It was a six-month standoff that immediately quadrupled the price of oil from $3 to more than $12 a barrel. Even after the embargo was formally called off, the world's economy never recovered from the shock of the global cartel and oil being used as a weapon.

In 1979, it was déjà vu all over again with the Islamic revolution in Iran. Between 1978 and 1981, oil prices more than doubled again, from $14 to $35 a barrel – and again the world was thrown into recession.

Oil was a weapon in itself, but even when oil production wasn't capriciously reduced by angry sheiks, the staggering cash flow it provided for Arab regimes allowed them to equip themselves with armed forces that would otherwise never have been available to them. And it drew the Soviet Union into the region, as both NATO and the Warsaw Pact carved up the neighbourhood and fought proxy wars through their local surrogates.

Israel isn't just like the Western democracies in its values. It became like the West in its dependence on energy imports, even though it was right in the middle of the Middle East.

Which meant Israel often had to buy its fuel from countries that at best were hostile – and occasionally were in an all-out war against it. In the late 1960s, Israel imported oil from Iran, when it was led by the shah; in 1979, the Ayatollah cut that off. More recently Israel imported natural gas from Egypt, a country with which it has had a cold peace since the late 1970s. But after the Arab Spring toppled Egypt's Hosni Mubarak, that natural gas pipeline was bombed fourteen times – something that the anti-Israel leadership of the Muslim Brotherhood hardly seemed to mind. They soon officially cancelled the pipeline contract, which had promised to supply Israel until at least 2028.[70]

What luck then that in early 2009,[71] a joint U.S.–Israeli consortium discovered a massive natural gas field just offshore Israel, called the Tamar field, with 10 trillion cubic feet of gas.[72] That field alone is expected to provide more than half of Israel's natural gas needs over the next decade. Add in other nearby undersea gas – like the massive Leviathan gas field within Israel's exclusive economic zone in the eastern Mediterranean Sea – and Israel now has enough natural gas to last for 150 years.[73] All told, that's close to 30 trillion cubic feet of gas; for comparison, it's more than ten times Germany's reserves.

The gas started flowing in the spring of 2013, through an underwater pipeline from the offshore rig. According to the Bank of Israel, the Tamar gas field alone is expected to tack on 1 per cent to Israel's GDP immediately, and more as the other offshore fields are developed.[74] Just the savings of using natural gas for Israeli power plants – instead of costlier fuel oil or diesel – is expected to save the country $14 million a day,[75] not an insignificant sum for a country of fewer than eight million people.

Israel's government has indicated that it will allow the gas consortium to export to other countries, a previously unthinkable circumstance. Israel could join its Arab rivals as an energy exporter. Nearby Cyprus not only has begun negotiations to import gas itself, but to serve as a large LNG facility to support shipping the gas to the far east.

Exporting gas isn't just economically beneficial, but it could have strategic benefits, too. One of the prospective importers of Israel's gas – the Muslim state of Turkey[76] – has frayed its relations with Israel over the past ten years. And Jordan, another Arab neighbour of Israel, has started negotiations[77] to import Israel's natural gas, as a more reliable source than Egypt.

The day the gas pipeline from the offshore rig started flowing, Silvan Shalom, Israel's minister of energy, wrote, "This is Israel's energy freedom and independence day."[78] It wasn't even about economics; it was about independence.

It's a job creator itself, and the price of electricity in Israel is expected to ease. But it's about no longer having to depend on neighbours who would just as soon bomb Israel as trade with it. "Gas flow will result in significant political and social change, and improve the standard of living and the environment," said Yitzhak Tshuva, the project's controlling shareholder.

There's an old Jewish joke: when Moses led the Israelites out of Egypt, why did he turn left, to resource-barren Israel? Why couldn't he have turned right, instead, to oil-rich Saudi Arabia? With Israel's new natural gas wealth, that joke just doesn't work any more.

QUEBEC

In the 1960s, fiercely nationalistic Quebecers launched a contemporary movement to declare independence from Canada. In 1967, Charles de Gaulle, the president of France, greeted a huge crowd from a balcony in Montreal (he had come to town ostensibly for the World Expo) and declared *"Vive le Quebec libre!"* – Long live a free Quebec! Quebec sovereigntists – those committed to making their Canadian province into an independent country – were elated. Their liberation from Canada was inevitable.

That was more than forty years ago, and Quebec has had a few moments nearly as climactic as that one. Voters have elected provincial governments that vowed to make Quebec independent once and for all. They sent separatist parties to the national government in Ottawa; the Bloc Québécois was once even the official opposition in Canada, the second most powerful party in Parliament, after the governing party. In 1980, the separatist Parti Québécois (PQ) provincial government held a referendum, asking voters for approval to separate. Back then, 40 per cent approved. In 1995, the PQ held another vote. This time 49.42 per cent did. The sovereigntists came within a hair's breadth of winning. But they lost.

In the meantime, Quebecers have mused about all the ways they would go it alone. About a new Quebec currency. A Quebec passport. But amid all this obsession about independence, Quebecers have remained entirely dependent on outside oil and gas. Not just Canadian oil and gas, in fact. Because of its proximity to the Atlantic, via the Saint Lawrence River, Quebec gets 92 per cent of its oil from overseas – from OPEC members like Algeria and Saudi Arabia.

Quebec imports far more foreign oil, as a percentage of usage, than even Americans do. And every single drop comes from outside the province. Every bit of natural gas, too. Quebec consumes a lot of fossil fuels every year: 500 million cubic feet a day – about as much as the U.S. states of Tennessee, Kentucky, and Montana combined. It gets cold in Quebec, after all. And Quebec uses roughly 400,000 barrels of oil every day, or about 140 million barrels a year – all of it purchased from somewhere outside Quebec. That's as much oil as they use in the whole country of Greece. Or Pakistan.

Quebecers are a long way off from being independent – but nowhere more than with their utter dependence on imported oil and gas. As a lot of nations have discovered, you can't chart your own national course when you're dependent on foreign imports for your most vital day-to-day energy needs, when you rely on foreign interests just to keep your economy powered. The United States knows a little bit about that – having to cozy up to Saudi Arabia, for instance, in order to ward off future oil shocks like the spiteful OPEC-designed oil crisis of 1973. But look at countries that have declared independence far more recently: former Soviet states like Bulgaria and Ukraine. They're at the mercy of Russia because Russia still controls their most precious commodity. Unlike those former Soviet states, Quebec could have begun producing its own oil years ago.

It has what are estimated to be billions of barrels offshore in the Gulf of St. Lawrence. But of course, the Quebec government imposed a moratorium on offshore drilling.[79] With Quebec's nearly 100 per cent reliance on imported fossil fuels, its independence, if it ever came, would be seriously compromised by its economic dependence on outside forces.

Smart sovereigntists know that. Lucien Bouchard was the former leader of the separatist party Bloc Québécois, when it was the federal opposition in Ottawa, and he was premier of Quebec, too, for five years, as leader of the separatist Parti Québécois. After that, though, Bouchard went on to become the head of the Quebec Oil and Gas Association – and he dedicated himself to convincing Quebecers that they needed to start tapping their own fossil fuel resources – through fracking – rather than chase independence while still wholly reliant on outsiders for one of their most crucial economic inputs.

That's not an easy mission. Quebecers already tend to lean further to the left than most Canadians, and they take their cues more from Paris, when it comes to environmental attitudes, than from Ottawa or Toronto. After France imposed a moratorium on shale-gas fracking, Quebec followed suit. Actually, it didn't just follow suit – it went off the deep end.

And yet, polling shows that 79 per cent of Quebecers want to be energy independent. Sixty-two per cent are against foreign energy imports. And a remarkable 84 per cent are in favour of developing natural gas locally. But 69

per cent of Quebecers are against using shale to get it –
even though they use shale gas, imported from outside the
province, daily.[80]

When it comes to its justice system, Quebec is famously
considered soft on crime, compared to the rest of the
country.[81] The largest opposition to reintroducing capital
punishment in Canada is in Quebec, by a significant
margin.[82] Quebec's provincial legislature voted unani-
mously in 2011 to demand changes to a sweeping federal
crime bill because it would have allowed adult sentences for
youths who commit especially serious violent crimes. Such
measures, said the declaration of Quebec's National
Assembly, "go against the interests of Quebec and Quebec
values as regards justice."[83] But Quebec's values regarding
justice for those involved in hydraulic fracturing are down-
right merciless. Anyone caught drilling, fracturing, or performing injectivity testing for shale gas in Quebec faces a
draconian penalty: fines from $1 million for individuals and
$6 million for companies, and a possible prison sentence of
up to three years.[84] Quebecers may have a more forgiving
attitude toward murderers, rapists, and child abductors, but
those who test for the existence of fuel are clearly inexcus-
able monsters.

And the Quebec government doesn't appear to believe
that will ever change. "I cannot see the day when the extraction of natural gas by the fracking method can be done in
a safe way," Martine Ouellet, Quebec's environment minister, said in 2012.[85] "We will impose a sweeping moratorium, both on exploration and on extraction of shale gas."

In early 2013, her government – a separatist Parti Québécois government – did just that, tabling a law that would bar any fracking activity or exploration in the St. Lawrence valley for the next five years.

It was a breathtakingly audacious sabotaging of Quebec's own interests – and Bouchard has made it clear he considers the anti-fracking impulse to be a woeful blow to Quebec's potential for energy independence. How could Quebec so readily dismiss technologies that even Barack Obama, the most environmentally obsessive president in American history, was backing as a justifiable means to achieve energy independence for the United States? How could Quebec completely dismiss an entire canon of studies that showed shale gas can be exploited safely, while creating jobs and government revenues?

"Americans aren't crazies. Our neighbours in English Canada aren't crazies. The Germans aren't crazies, nor are the Chinese," Bouchard said. "Let's see what's being done elsewhere and listen to the experts. And I think we should hear our own experts and let them finish their work."[86]

Because Quebec isn't just dependent on outside interests for its oil and gas; it's dependent on others for its fiscal solvency. With a $250-billion debt, Quebecers are deeper in hock, on a per-capita basis, than even Spain or Portugal. And that's after receiving nearly $8 billion a year in forced "equalization" transfers from richer provinces. That's money that, under the Canadian constitution, richer provinces have to share with provinces that have lower revenues – even if those "have not" provinces could be earning more money by doing things like exploiting their own resources, as Quebec could be doing.

More than 10 per cent of Quebec's annual budget is paid for by better-managed provinces like Alberta, Saskatchewan, and British Columbia. Of course, those provinces get a significant amount of their revenue from oil and gas. So Quebec already depends largely on oil and gas revenues – from other provinces. From fracking in other provinces. And it depends on gas imports from other provinces – shale gas that was fracked in other provinces. It just refuses to tap any of its own gas, or gas revenues.

But there is an estimated 155 trillion cubic feet of shale gas in Quebec – enough to supply Quebecers with their current annual demand for more than seven hundred years. If a fifth of that is, as projected, recoverable using current fracking technologies, that's still a century-and-a-half's worth of Quebec's current gas consumption. And at about $3.90 per thousand cubic feet (the wholesale U.S. market price for pipeline gas in the first quarter of 2013), Quebec's 155 trillion cubic feet of shale gas would be worth more than $600 billion. Even if no more than a fifth of it is ever recovered, if fracking technologies never improve – as improbable as that is given the consistent track record of technological advances – that's $120 billion worth of shale gas, enough to pay off half of a provincial debt that took decades to rack up.

And then there's all the oil that Quebec has, too. Even though the province doesn't produce a drop, geologists long ago identified offshore oil in the Gaspé Peninsula, the Gulf of Saint Lawrence, and Anticosti Island. The reserves are considered to be significant but have yet to be fully assessed because, again, Quebec refuses to entertain the idea of off-shore drilling. Meanwhile, one U.S.-registered exploration firm is launching a challenge under the North American Free Trade Agreement (NAFTA), after it saw its permits to explore for shale gas in the St. Lawrence River negated by the government's announced fracking moratorium.[87] Lone Pine Resources points out that it received no compensation from the province – which was more than happy to sell the permits in the first place – and wants $250 million in redress. If Lone Pine wins, and it certainly seems to have a fair case, it will be the federal government that pays up, under NAFTA rules, and Canadian taxpayers paying yet again for Quebec's energy inanity.

It's possible that Quebec's hostility to energy extraction is due to something as basic as a language barrier. Nearly 80 per cent of Quebecers are francophone. That number rises to 95 per cent in Quebec City, the capital. In a country that is, outside Quebec, overwhelmingly English-speaking (only 20 per cent of Canadians claim French as their first lan-guage and 92 per cent of those live in Quebec), Quebecers live in a kind of linguistic bubble inside Canada. They're more likely to get their wider news coverage from France – the heartland of anti-fracking hysteria – than from Canada or the United States. They get the Parisian media's perspec-tive on fossil fuels, not the more normal version from prov-inces and states where gas extraction is commonplace and where fracking is having an immensely positive impact.

Ironically, Bouchard has pointed out that many nation-alistic Quebecers are determined not to let non-Quebec gas companies come in and pump gas that belongs to the province. "The view is that these people are predators and they're stealing our resources," Bouchard has said. "It's extraordinary what's happening."[88]

Most resource-rich jurisdictions aren't nearly that para-noid – or at least, they know they can profit just as easily from foreign operators. In Texas, California, Alberta, or Britain, foreign companies are free to invest in oil and gas, as long as they pay the mineral owner (usually the govern-ment) a healthy royalty. But instead of profiting from explorers from outside Quebec, or even developing their own homegrown gas industry to tap that vast shale resource, Quebecers have just shut it down. In 2012, Quebec even established a specialized school specifically to train workers for the oil and gas industry:[89] what a shame that none of them will be able to work in their home province. The pro-vincial leaders who have for decades claimed to want nothing more than independence for their people have now consigned Quebecers to even deeper energy depend-ence for years.

NEW BRUNSWICK

For more than a century, New Brunswick – the Canadian province that borders Maine to the west and the Gulf of Saint Lawrence to the east – has been what they call in Canada a "have not" province. Less politely put, it's a province dependent on constitutionally mandated "equalization" payments from more financially secure provinces just to keep up with the national Canadian standard of living. Even less politely put, New Brunswick is poor. It's the third-poorest province in Canada. Its per-capita GDP is 18 per cent lower than the Canadian average. But compared with an energy-producing province, like Alberta or Saskatchewan, its GDP per person is more than 40 per cent lower.

New Brunswick governments have tried almost everything to goose economic growth in their province. In the seventies, taxpayers shovelled subsidies into the Bricklin Motors company, with visions of starting a Canadian-made car company. The Bricklin SV-1, designed by the same guy who helped create TV's Batmobile and backed by Malcolm Bricklin, who founded Subaru's American operation, was sleek, with gull-wing doors, but it was a huge money-loser. Bricklin spent $16,000 to make each SV-1 and then sold them to dealers for $5,000. Fewer than 3,000 cars were made before Bricklin went out of business, taking millions in taxpayer loans with it.

In the nineties, the provincial government had a grand plan to use subsidies to lure call-centre operations – for banks, tech companies, travel agencies, etc. – to New Brunswick. The province would be the call-centre centre of North America. That must have sounded like a plausible plan – until India decided to turn itself into the call-centre centre of the world. New Brunswickers might need the work – the province's 10.5 per cent unemployment rate in the first quarter of 2013 was double that of oil-rich Alberta's

– but with a legislated minimum wage of $10 an hour, they weren't going to stand much of a chance competing with New Delhi.

The next big idea to hit New Brunswick was natural gas. Not just drilling for gas: with 10 million cubic feet of gas produced each day, N.B. wasn't much of a producer. Nearby Nova Scotia was producing ten times that, never mind a more energy-rich province like British Columbia, which produces 3.5 billion cubic feet a day. New Brunswick was going to import gas, from abroad. Liquefied natural gas (LNG).

Canaport, a facility designed to accept liquefied natural gas from Qatar and other gas-producing nations and re-gasify it, started construction in the mid-2000s. It accepted its first shipment in 2009. When Canaport was about to open, it must have been a very hopeful time for the province. Co-owned by Spanish-based Repsol and locally based refining firm Irving Oil, Canaport was the first LNG facility to be built on the North American east coast in thirty years. By the time it was operational, a lot of that hope must have been dissipating. The timing couldn't have been worse.

In July 2008, the U.S. wellhead price for natural gas had hit a record high – $10.79 per thousand cubic feet. By June 2009, just eleven months later, when that tanker with three billion cubic feet of LNG pulled in from Trinidad and Tobago, prices had collapsed by more than two-thirds, to $3.38 per thousand cubic feet. And they would never again climb anywhere near to where they had been.

Canaport today is running well below capacity – using just 30 per cent of its capacity. By early 2013, with gas still trading at around US$4, Repsol had decided to bail on the natural gas business and on Canaport. Its debt rating was in danger of being downgraded to junk status. But it was Canaport – and a 25-year commitment to ship gas into Canada – that blocked the deal. Nobody wanted to be saddled with an LNG-import facility on the east coast, when northeastern states like New York were suddenly awash with a glut of shale gas. Canaport had "become a big white elephant for Repsol due to the North American gas-supply demand fundamentals," Cameron Gingrich, senior manager of gas services at Ziff Energy Group, told Bloomberg in February 2013.[90] Canaport wasn't alone: all LNG import operations in New England were running idle, or at least close to it.

With all that gas suddenly coming online in the northeast United States, who would bother with the added cost of liquefying gas from Qatar, or even the Caribbean, shipping it across the ocean, and then re-gasifying it? It simply made no economic sense, and no LNG operation could possibly sell at the kind of price to justify it. By March 2013, Repsol had to write down $1.3 billion in lost value on Canaport, after selling off other LNG assets, outside the United States and Canada, to Shell. Shell didn't want Canaport, because low regional gas prices meant the terminal could not be "adequately valued," Repsol said in a statement. Given that a Repsol fact sheet on the project said that "Canaport LNG's founding partners invested approximately

$1.2 billion to develop the Terminal,"[91] it appeared that Repsol was essentially declaring the New Brunswick terminal virtually worthless.

"It's pretty hard to make a case for bringing LNG in anywhere in North America when we have such a supply of unconventional gas that's been brought on," Ed Kallio, director of gas consulting at Ziff Energy Group, told Canada's *Financial Post*. "Since 2008, we've added 15 billion cubic feet per day of supply in the Lower 48. A lot of supply has come on and it's backed out LNG."[92]

Canaport could, theoretically, be converted to an export facility – at the painful cost of $2.5 billion to $4 billion, according to some estimates. But that would require a massive, reliable supply of natural gas, and the pipeline infrastructure from the United States that could provide that kind of volume to New Brunswick just does not exist. New Brunswick, it seems, is stuck with yet one more failure in its attempt to become an economic force.

But New Brunswick may finally have something that the world would dearly value – not uneconomical sports cars, overpriced call centres, or foreign liquefied natural gas imports. It has shale – the thickest shale gas reservoir in North America is located in New Brunswick.

Nobody knows yet how much gas is trapped in that thick reservoir; the exploration is just beginning. But Frank McKenna, the province's former premier and a former Canadian ambassador to the United States, has estimated that developing New Brunswick's shale-gas industry could generate more than $7 billion in royalties and tax revenues for the government. For a province with an annual budget just over $7 billion, that's an incredible amount of money. Imagine if California had suddenly stumbled on a new industry worth $100 billion – the total of its annual state budget.

That's why the New Brunswick government is taking shale gas very seriously. All that potential for shale gas combined with an Atlantic seaport – with the potential for exporting LNG to Asia or Europe, replacing the now-uneconomic import plan – could feasibly turn New Brunswick into a vibrant energy-export hub someday. The distance for a ship to travel from New Brunswick to energy-hungry India is shorter than it would be for a ship leaving from Canada's west coast. A couple of hundred years ago, Saint John, New Brunswick, was one of North America's most dynamic ports, competing with Boston and New Haven for exports, like lumber, to Great Britain, and with a thriving shipbuilding industry. New Brunswick has the right location for global trade; it just no longer has a product that the rest of the world wants. But there are hundreds of millions of people around the world who could use more gas.

In May 2013, the provincial government released a "blueprint" for energy exploration in the province. The premier, David Alward, called it a "key part of our government's plan to rebuild New Brunswick's economy and create jobs here at home."[93] The blueprint considers how to balance the interests of water management, air quality, royalty regimes, workforce development, economic development and attracting investment, supply planning, and opportunities to get economically challenged First Nations bands involved in the energy industry. It has already granted an exploration licence to SWN Resources Canada to search 2.5 million acres of the province for gas, and SWN has committed to investing a minimum of $47 million into the project.

Alward is taking an exceedingly careful and gradual approach to rolling out shale-gas development. He should. He needs to. New Brunswick borders Quebec, where anti-fracking hysteria has reached absurd proportions, with the Quebec government imposing a draconian moratorium on shale-gas exploration, as we've seen. New Brunswick is also close to New England, where the American anti-fracking movement has managed a firm foothold. Environmentalist groups have already started working to mobilize Indian bands in New Brunswick, exploiting them to hijack aboriginal interests just as they have done in western Canada and Quebec. Anti-fracking groups affiliated with *Gasland* director Josh Fox have already begun moving in.

And they're stirring up trouble, using fear. In June 2013, members of the Elsipogtog Indian Nation band confronted a seismic testing crew from SWN Resources, accompanied by protestors claiming to be "independent UN observers" (there are no UN observers in Canada). After a heated encounter, the demonstrators towed the company's trucks onto the nearby native reserves – essentially stealing millions of dollars' worth of equipment. Later that same month, police arrested protestors trying to block SWN trucks from doing testing in the same area.[94] The protests turned even more violent and destructive when protestors turned to arson, setting ablaze two of SWN's seismic rigs.[95] Police had to arrest more than thirty people for breaking the law in the onslaught against the company and its employees.

This is what New Brunswick is up against: the opportunity to develop its economy, finally, in a sustainable and market-oriented way, facing resistance from a handful of professional anti-fracking agitators using local native groups as their pawns. It won't be an easy battle – the anti-fracking lobby has a lot of resources and support behind them, from well beyond New Brunswick. But New Brunswick has known for too long what it means to be dependent on others just to get by. To the province's credit, it seems, so far, determined to overcome the attacks by anti-fracking groups and finally build energy security, and a thriving economy, for itself.

NOVA SCOTIA

There are few Canadian provinces with economies that are more depressed than New Brunswick's – but Nova Scotia is one of them. New Brunswick may be struggling, but it can at least boast that it has a per-capita GDP that's nearly 10 per cent higher than its Atlantic neighbour, Nova Scotia.

Naturally that means enterprising young people don't have much reason to stick around Nova Scotia. Not with a nearly 9 per cent unemployment rate province-wide, and over 16 per cent in some parts of the province. This isn't some slow recovery from the global recession, either. From 2000 to 2010 Nova Scotia's average unemployment rate was 8.8 per cent. What would pass for distressing job numbers most anywhere else are the standard in Nova Scotia, even when the rest of the continent was booming.

Young people who want to work, who want security, who want a future, aren't staying in Nova Scotia. They're leaving home – and a lot of them, naturally, are travelling to the west: Alberta and B.C., where there is a high demand for workers. The reason, of course, is that those provinces have something in high global demand: fossil fuels. People who have made the move, and there are many of them, know that if you have a skilled trade, like welding, you can make double the hourly wage in Alberta than you'll make in Nova Scotia. Even if you don't have a skill, just working in a coffee shop can make you 50 per cent more per hour in Calgary than you'll earn in Saint John (which is, most likely, the minimum legal wage).

What does a province without a growing economy look like? Old. And disproportionately female. In 1982, Statistics Canada shows that the median age of a N.S. resident was about 29 years. Today, it's 43 – and getting older by the day, as young people decamp for more promising economic opportunities elsewhere. Thirty years ago, the ratio of males to females in N.S. was right around the Canadian average – roughly balanced (with a slightly higher number of females, which is nature's way of accounting for greater mortality rates among males). Today, that ratio is completely out of whack, with just 94 males for every 100 females – the most unbalanced sex ratio of all 10 provinces. The majority of those people who are leaving N.S. for work is, of course, men. This is what a province in decline looks like, shrivelling up for lack of an economic future.

Well, more accurately, Nova Scotia has an economic future, but its political leaders have so far refused to take advantage of it. The province is sitting on the Horton Bluff Shale deposit, containing an estimated 17 trillion cubic feet of natural gas, 3.4 trillion of which is recoverable with current technology. That's a bigger recoverable reserve than in Saskatchewan and Manitoba combined. Like New Brunswick, Nova Scotia's Atlantic coastal location means it's in a position to export natural gas to the world.

Unlike in New Brunswick, the government in Nova Scotia appears uninterested in even considering such a historic economic-development opportunity. In 2008, shale-gas developer Triangle Petroleum set up shop in Nova Scotia to begin testing for shale-gas drilling sites. The initial findings were very encouraging, which isn't surprising given the amount of shale gas already estimated to be in the Horton Bluff. But in early 2011, the province put a halt to all that potential, announcing it wanted to freeze all shale-gas exploration and development, appointing a committee, instead, to "examine the environmental issues associated" with fracking and to "determine how they are managed in other jurisdictions and identify industry best practices." No shale-gas exploration or production would be allowed while the committee did its work.

There's nothing wrong with researching environmental issues. But when it comes to "industry best practices" in drilling, that's usually something industry is a lot better at worrying about than government is. That's why they're called industry best practices, and not government best practices. But even if Nova Scotia is determined to expand its regulatory scope from environmental protection into the area of industry practices, it doesn't seem in any great hurry to figure it out.

This fact-finding moratorium has been in place for more than two years. Jurisdictions all over North America are pumping shale gas out in enormous quantities. There is no shortage of jurisdictions to "examine"; in some parts of North America, fracking has been going on for decades and there is already an abundant volume of information for Nova Scotia's regulators to examine. Yet, in 2012, Nova Scotia decided it still needed much more time and extended the review to "mid 2014" to "ensure it has the best information to make decisions." And, of course, the moratorium was extended, too. Who knows if, come mid-2014, the committee decides it needs yet more time to ponder a practice already underway in literally thousands of places across North America, and more overseas? In a province with nearly 9 per cent unemployment, getting paid to sit on a review committee has got to be a rare privilege, but the longer the committee sits around thinking about fracking, the longer the rest of the working-age population sits around waiting for some economic growth. Or, more likely, they won't wait around – they'll just move, as thousands of Nova Scotians already have, to provinces more eager to seize an economic opportunity when they see one.

Chapter Twelve

GASLAND AND JOSH FOX

Given how energy-intensive moviemaking is, Hollywood has a strange hatred for the fossil fuels. James Cameron, the moviemaker behind such blockbusters as *Titanic* and the Aliens series, flew to 107 different cities to promote his movie *Avatar*[1] – a film about the evils of overconsuming natural resources. The environmentalist mantra of "reduce, reuse, recycle" is the opposite of Hollywood materialism, especially for multimillionaires like Cameron.

But Hollywood isn't about reality – it's about perception, even about fantasy. It's why stars like Leonardo DiCaprio fly in private jets, but then make sure the paparazzi see them pulling up to a charity fundraiser driving a Toyota Prius. It's show business.

And for many actors – especially those like Matt Damon, who has been the butt of jokes for being rather dim-witted (in the satirical movie *Team America*, the only line the Matt Damon character ever says is "Matt Damon") – being politically fashionable can be a way to seem serious and thoughtful, not just another pretty face reading lines written by someone else.

Becoming a political protestor is a smart career move for other Hollywood stars, for a different reason: if their acting or singing career is past its prime, they still can milk whatever lingering name recognition they have by being an outspoken social activist.

Of course, there are plenty of dire social problems in need of celebrity activists – there has been no shortage of wars or terrorism to oppose and other grave causes. But for some reason Hollywood just seems braver when it's attacking American industries where there's no risk their films will be banned or they'll be arrested.

Darryl Hannah, the starlet most famous for the 1984 film *Splash*, hasn't had a film hit in over ten years, since her cameo appearance in *Kill Bill*. But she has found a new lease on life – or at least publicity – by demonstrating against mining in West Virginia and the Keystone XL oil pipeline from Canada, getting arrested at both demonstrations.

Matt Damon still gets starring roles in Hollywood, so he doesn't have to use his body as a political weapon. But just like Darryl Hannah, he seems desperate to show the world there's more to him than just a pretty face.

So in December 2012, the movie *Promised Land* premiered, starring Damon, who also co-wrote the screenplay. It's a bizarre, conspiracy-theory–filled anti-fracking movie in which Damon plays a fracking company employee who eventually turns against fracking because of shockingly dishonest tactics by his employer.

The film was a total bust, grossing less than \$8 million[2] in the United States and being pulled from theatres after just four weeks. But if a million Americans saw Matt Damon, the man they love and trust from the Bourne movies, show them just how dangerous and dishonest fracking companies are, that's a lot of hearts and minds turned against the industry.

What those one million moviegoers likely didn't know, though, was that Damon's film was financed in part by Image Nation, a company owned by the dictatorship of the United Arab Emirates, an OPEC country that also has the world's seventh-largest reserves[3] of natural gas. Fracking is a direct threat to their market share and the price they get per cubic foot of gas – and an anti-fracking propaganda film like Promised Land might slow down fracking just a bit, or even tip some jurisdictions against the technology.

When Phelim McAleer, the filmmaker behind the pro-fracking film Frack Nation, confronted Damon about it,[4] Damon laughed at first; then challenged McAleer about his own film; and then finally claimed that he, as the star, screenwriter, and producer, had no idea until the filming was over that his own film received the OPEC money. And as a great dramatic actor, he said that to McAleer with a straight face.

But there are plenty of celebrities, or ex-celebrities, more than willing to attack fracking without OPEC funding.

Yoko Ono, the widow of John Lennon, joined her son Sean Lennon and activist emeritus Susan Sarandon for a luxury tour of Pennsylvania counties, stopping in to visit anti-fracking activists, TV cameras in tow. The tour was organized by Artists Against Fracking run by Yoko and Sean. And it surely was a coincidence that the publicity tour happened to be timed for the same week Sean released his new album.

Fracking is complicated; it sounds alien; it takes a while to explain. For low-information voters – the kind of people who would listen to a celebrity over a scientist – the emotional appeal of Artists Against Fracking would likely be all they'd hear on the subject.

But of course the Artists didn't go from New York to Pennsylvania on bicycles. It was the winter, with snow on the ground. They rode in a luxury Mercedes bus, nicely heated in the winter. They used plenty of fossil fuels to get there, and their luxury apartments back in Manhattan surely had natural gas in their furnaces and their stovetops too.

If not fracked gas, then what did they prefer? One of their entourage, Arun Gandhi, one of Mohandas Gandhi's grandsons, told reporters that any form of fossil fuels – such as the one they had just driven in on – was "violence against nature, against resources, against environment. Eventually this is going to destroy us."[5]

Why did he make that absurd comment? But more importantly, why is the grandson of the founder of modern India – the man who stared down the British Empire, who used passive resistance and hunger strikes to help give his great country independence – why is Gandhi's grandson so unbusy that the greatest human-rights campaign he chooses to busy himself with is a group of dilettantes bussing in to small-town Pennsylvania to oppose a kind of drilling that's been done a million times across America?

Is fighting against fracking, in one of the freest countries in the world, truly the most important human-rights calling for the great Gandhi's grandson? Is every other challenge in the world licked, every other problem solved?

Or, like Sean Lennon, was Arun Gandhi just mugging for the cameras, just trying to get a little free PR for his next gig, whatever that may be?

Fracking is an enormous industry. But anti-fracking has become a pretty lucrative business too. It's the business of opposing fracking – and it's an industry full of lawyers, lobbyists, PR men, and celebrity endorsers. They've done in a few short years what the industry itself didn't do in fifty: they made it famous. Sort of like how the Japanese Imperial Navy made Pearl Harbor famous.

The chief entrepreneur in the growth industry of anti-fracking activism is Josh Fox, a New York filmmaker known for his carefully crafted hipster image: Buddy Holly glasses, a *Miami Vice*–style three-day growth of beard, and when all else fails, self-consciously plucking a banjo.

Fox attended an Ivy League school – New York's Columbia University – and then puttered around the New York theatrical scene.[6] Until he discovered fracking, he was just another struggling artist, but one with a dramatic flair that stood out even in an industry full of peacocks.

The New York Times spotted his showmanship in 2004, when they called him "one of the most adventurous impresarios of the New York avant-garde." Time Out NY said he was "one of downtown's most audacious auteurs," with a "brilliantly resourceful mastery of stagecraft."[7] He must have always known he was destined for the big time; he founded a company with the modest name of International wow Co., headquartered in Brooklyn.[8]

International wow's first feature film[9] was called Memorial Day, about the U.S. holiday. Fox says, "Memorial Day weekend . . . is a time for remembering war and a time to party your ass off and get totally shitfaced." That might come as news to millions of American veterans who remember history's wars with solemn wreath-laying ceremonies at veterans halls. But Fox says the holiday is about the sexualization of violence and the grooming of a next generation of fighters: "Violent emotions – sexual and bloody – Memorial Day becomes a day to hit the beach and get rocked and throw up on your date. . . . Those partying drunk naked kids ARE thinking about war, they just don't know it." Fox wouldn't be the first filmmaker in history to use his art as a form of personal psychotherapy or to project his own views on the whole country. That's why it's called art. But even back when he was making unwatchable indulgences, an anti-American theme started to appear.

Fox became a founding member of THAW – Theaters Against War – a group of hard-left activists[10] who hosted anti-war rallies and sent money to Palestinian activist groups. They even hosted panel discussions by family members, defending accused terrorists. Fox followed their lead with his next project, an interactive play[11] in which audience members were instructed to dress up as U.S. soldiers and to join the actors on stage, going through scenes shooting people – including innocent civilians. "This piece is completely apolitical," Fox told one reviewer with a straight face. But he couldn't keep that up for long. "I'm standing here in the middle, trying to be a provocateur," he said.

But writing anti-American, anti-war shows was pretty standard fare for angry young filmmakers in New York during the last years of the Bush administration. Fox needed to break out with something fresh.

It was in his jihad against fracking that Fox has found his calling, truly creating an international "wow." Making *Gasland* in 2010, an advocacy film against fracking, turned Fox into a true celebrity, even earning him an Oscar nomination. It was like Michael Moore's breakthrough agitprop film, *Roger and Me*.

Like Moore did in his films, Fox cast himself as the star of *Gasland*. Without any scientific or environmental credentials, his story needed another source of credibility, so Fox portrayed himself as someone who was nearly victimized by fracking, someone whom natural gas drillers tried to bribe with a huge cheque for permission to frack his property. That's the narrative arc of Fox's movie; that's the source of his legitimacy; it's his official explanation for why he started his campaign. Political campaign managers call that a "back story": who were you before you became a politician? Fox presents himself as a just-folks guy who started to research fracking only when he was offered a deal. The film purports to document his journey toward the truth – as a cautionary tale for others who are curious about fracking. And you know he's being honest about his motivations, because who else would turn down an offer from a drilling company of nearly $100,000?

Except it's just not true. Fox did in fact grow up in Pennsylvania. Presumably that's where he learned how to pluck a banjo, Louisiana-style. But he's lived and worked in New York for decades. It's true that his family[12] has a weekend property in Pennsylvania, but neither of his parents or his siblings live there. It's what other New Yorkers might call a vacation cottage. But Fox claims it's his primary residence – even though in real life he lives in New York.

In the opening scenes of *Gasland*, he's sitting on his deck in the beautiful Pennsylvania wilderness – a million miles from places where they use words like "downtown's most audacious auteurs" – brandishing what he claims to be the offer from a gas company, with the company's name blacked out.

Except it wasn't.[13] The reason the fracking company's name is blacked out in *Gasland* is that the offer did not come from a gas company. Though Fox flashes the contract offer on the screen for only a few seconds, it was enough for Marian Schweighofer of Northern Wayne Property Owners Alliance[14] to identify it as her group's standard contract. It was her landowner group's suggested contract for any of its members as a counter-offer to a gas company.

When an interviewer on National Public Radio – a liberal-leaning network normally sympathetic to anti-fracking activists – brought up this inconsistency to Fox in an interview, he evaded it several times and then asked if he could go "off the record" to answer it.

The anti-fracking arguments made by Fox can stand or fall on their own merits. Whether fracking poisons the water or air is an objective fact that can be measured empirically – it doesn't depend on Fox's identity and motivations. But if the central premise of the movie – that Fox didn't choose this fight, it chose him; that he was a disinterested Pennsylvanian, just enjoying the rustic beauty of the countryside, until the fracking companies came calling – if that whole back story is false . . . then why the movie? Is there an alternative rationale for making his film that he wants to downplay? Does he have an ulterior motive that he wants to disguise? Is he a neutral documentary filmmaker, or even a passionate advocate, but one who lays all his cards on the table? Or is there something else going on?

Josh Fox has meticulously groomed his image as a young independent filmmaker who was motivated by an honest desire to find the truth. But his little independent film has apparently caught the admiration of the Kremlin-controlled Russian gas company Gazprom.[15] And after the Academy Awards chose another nominee as the best documentary, the Venezuelan Embassy in Washington, D.C., publicly tweeted, "Sadly, 'Gasland' didn't win an Oscar, because a

Vzlan helped make it."[16] Of course, it wasn't just any Venezuelan. It was an employee of the government's state-run film office that Fox even gives a shout-out to in the film's credits.[17]

Why did an OPEC country with an anti-American foreign policy decide to support a movie attacking America's domestic energy industry? Is it because fracking is also used to produce enormous amounts of crude oil in the United States, reducing Venezuela's daily exports to the United States by nearly 10 per cent in the past year alone?[18]

Is that why Fox goes to such lengths to establish a back story as a good old American woodsman, a hardy patriot who practically bleeds red, white, and blue?

Fox's movie should stand on its facts, not his biography or the admiration of foreign governments. But if he takes liberties with the facts of his own biography, does he take liberty with other facts in the movie?

And – just as troubling – what does it say about the objectivity of the media and film establishment that has lavished praise (and an Oscar nomination) on Fox's film, so uncritically, for so long?

All great filmmakers have an ego; it takes a confidence bordering on hubris to think that thousands or even millions of people should listen to you for an hour and adopt your world view. There's nothing wrong with a filmmaker wanting to be a celebrity. But how is faking a documentary any different than other famous hoaxes, like Stephen Glass's fabrications at the New Republic or Jayson Blair's at the New York Times that got those two men fired?

Sixteen minutes into *Gasland*, Fox says, "Apparently they were buying this act of me being a documentary film-maker." To most of *Gasland*'s viewers, that would be an endearing moment of self-deprecation. But maybe it was more – a subconscious confession that the audacious auteur with a mastery of stagecraft and a constantly shifting home address is getting away with it.

But still; even a trickster can be right on the facts. So, is he?

Five minutes into the movie, Fox claims that in 2005, the U.S. government exempted the oil and gas industry from water safety laws. Fox calls it the Halliburton amendment and wasn't shy about making the link to then vice-president Dick Cheney, a former Halliburton executive who remains a Darth Vader–like figure to the Left. Well, is it true? Did Cheney and George W. Bush – another former oil and gas executive – sneak a fracking exemption into the Safe Drinking Water Act? Was it "pushed through Congress by Dick Cheney"?

There was indeed a law called the Energy Policy Act of 2005, and it did indeed amend the Safe Drinking Water Act. But that law had never regulated fracking in the first place. The law made no change to the regulation of fracking, which is primarily regulated at the state level. The Environmental Protection Agency's (EPA) own website[19] makes this crystal clear: the EPA's Underground Injection Council does regulate any fracking done with diesel fuel. And "state oil and gas agencies may have additional regulations for hydraulic fracturing. In addition, states or EPA have authority under the Clean Water Act to regulate discharge of produced waters from hydraulic fracturing operations." That's how it was done before 2005, and how it's done now.

No doubt Dick Cheney supported the bill – he helped craft the government's energy policy, and his boss signed it into law. But it's not the White House that passes laws – that's Congress's job, and the bill had broad support from both parties. The U.S. Senate voted 74–26 in favour of the Energy Policy Act, and among those 74 supporters was the junior senator from Illinois at the time. His name was Barack Obama.

As we've seen, fracking happens hundreds, and usually thousands, of feet below the water table – no water wells go down a mile or two deep, where fracking usually happens. It doesn't make sense to regulate fracking in a drinking water law. But if a driller does, by chance, pollute a reservoir that people use for drinking, the EPA and state regulators have the power to stop the drilling and punish the company.[20] That, too, remains unchanged by the 2005 law.

In the main, the EPA doesn't regulate fracking – that's done by the states. But the EPA is constantly studying fracking, reviewing and compiling research done by other agencies, and conducting its own scientific inquiries into the safety of the process.

Is it important that Fox tried to paint a bipartisan U.S. energy bill as a Republican project, pushed through by Dick Cheney? No; but it is important that Fox leaves the distinct impression that fracking is unregulated, and that drinking water was stripped of its protection. That's not a partisan enthusiasm. That's just untrue. But it's also utterly essential to *Gasland*'s official narrative: a young, disinterested Pennsylvanian banjo player was almost tricked into signing a rich fracking lease; thank God he didn't because that stuff's completely unregulated. And in return for taking the big cheque, our hero would have been subjected to 596 chemicals — most of them toxic, many of them cancer-causing.

There are hundreds of different chemicals that can be used in fracking. In fact, the U.S. Congress added up more than 700 of them, used by 14 different companies.[21] And even though Fox lists many of them in his film, he says they're secret. Which is true in the same way that Coca-Cola's recipe is a corporate secret – the ingredients of Coca-Cola are printed on the side of every can, but the exact combination is private. But even Coke's claims to privacy don't work in fracking: in many U.S. states and Canadian provinces, fracking companies are required to list the exact chemicals being used on any particular drilling site – typically about a dozen. A *New York Times* review of Wyoming's publicly disclosed frack jobs showed between eight and fourteen chemicals being used for any one drilling site.[22] Anyone in the world – landowners, journalists, competitors – can see what's being used, corporate secrets be damned.

Still, that's exciting stuff for someone looking to paint a picture of fear. Even if it becomes less exciting when you learn that the mixture of these chemical additives typically makes up between 0.5 per cent and 2 per cent of frack fluid, the other 99 per cent being water and sand.

But Fox knows it's easy to scare people with the use of the word "chemical," even if it's a dozen of them, not hundreds, and even if they make up just half a percent of what's in fracking fluid. Because chemicals sound dangerous and mysterious.

Of course, everything is a chemical – that's just what we call different combinations of naturally occurring elements. Oxygen is a chemical. Chlorine sounds terrifying, but it's the building block of table salt. It can have a toxic effect on life, which is precisely why chlorine is so important to purify our drinking water. But it all depends on what story you're trying to tell. And in Josh Fox's case, it's pretty clear he wants to make chemicals sound as dangerous as possible. No doubt if he were doing an exposé on salt water, he'd talk about the toxicity of evil chlorine, and the mysterious dihydrogen monoxide – so hard to pronounce! Because that is the technical name of H_2O, or as most folks from Pennsylvania would call it, "water."

The chemicals are needed in fracking for a simple reason: they help the other 99 per cent of the fracking fluid – the water and the sand – get into all the little crevices of the shale rock, thousands of feet beneath the surface. That's why the particular blend used for each well is slightly different – it depends on the geology.

Fox works hard to highlight the scariest names of chemicals in his film. The others he flashes across the screen so quickly you have to stop the film to read them. One fracking chemical that whizzes by on screen is called guar gum. Perhaps Fox doesn't linger on it because it's more

commonly found in ice cream. It's an emulsifier – a chemical that keeps different things mixed together. Others of
these "toxic chemicals" include ground pecan shells and
ground walnut shells. Staying with the g's on his list, there's
graphite (used in everything from pencil lead to brake
linings); gypsum (used in plaster casts and plaster walls);
glycerol (a sweetener used in foods and medicine); and
galactomannan (used to make cream cheese).

No one is going to claim that fracking fluid is delicious,
or that all the additives are benign – though Halliburton
executives have been known to drink fracking fluid in
public,[23] as the ultimate testament to non-toxicity.

It's true fracking fluid often has other, less delicious
chemicals in it. Like ethylbenzene. That's one of the terrifying chemicals that Fox slows down and emphasizes, with
the full-screen declaration that it's a "known carcinogen."

Except that's not true. The EPA has an entire website[24]
dedicated to ethylbenzene. It's not something you'd want to
bathe in – nor would you have a shower in gasoline or many
other household chemicals. But it's crystal clear: "EPA has
classified ethylbenzene as a Group D, not classifiable as to
human carcinogenicity." They go on to say, "The only available human cancer study monitored the conditions of
workers exposed to ethylbenzene for 10 years, with no
tumors reported." But Fox wouldn't let that little detail get
in the way of his work. That's the knack for "theatricality"
the New York Times loves about him.

Which is good: ethylbenzene is one of the most important chemicals in our everyday lives – the vast majority of it

being used to make polystyrene, a.k.a. plastic. It's found in plastic cutlery and the plastic serving containers for individual yogurt and pudding cups. It's in road asphalt too. You wouldn't eat it – just like you wouldn't eat a plastic food container. But it's not going to give you cancer.

In addition to ethylbenzene, Fox highlights another unpronounceable chemical name on the screen: 1-(Thiocyanomethylthio)benzothiazole. Any chemical with numbers and parentheses in its name has got to be terrifying – and deadly.

Except that chemical – known as TCMTB[25] for short – isn't used to make us sick. It's used to keep us healthy. It's a fungicide, used in everything from carpets to leather goods to sewage treatment plants. You wouldn't want to gargle with the stuff. But no one does. It's used in a controlled manner, to keep us safe. It's been regulated and approved by the U.S. government since 1980.

Our food isn't soaked in TCMTB, but our food seeds are – according to the EPA, TCMTB is appropriate to keep harmful fungus off seeds for plants such as wheat, oats, and cotton.

TCMTB has nothing to do with oil and gas or fracking. The only reason it's one of the hundreds of chemicals on the list of fracking additives is that it's sometimes used if there is a fungus problem. Blaming TCMTB on fracking is as nonsensical as blaming it on bakeries, whose bread was made from wheat whose seeds were sprayed with the stuff. But science isn't Josh Fox's strong suit – or even honesty. Fearmongering is, and nothing says scary like a chemical whose name is thirty-three letters and numbers long.

In the first dozen minutes of the 101-minute film, Fox has done what he needs to do: scare viewers in a blizzard of legalese and scientific jargon – enough that most non-expert viewers would trust that Fox himself had done his homework on the facts so they didn't have to. They could trust him that fracking was unregulated and environmentally dangerous. So for the rest of the film, Fox could use his theatrical skills to do his real work: leave an emotional impact, using shocking personal stories, of how devastating fracking can be. With the scientific and legal preamble out of the way, viewers can trust not only Fox, but the people he interviews. They seem sympathetic, hard done by, and credible, not like cranks of conspiracy theorists. Why would this nice filmmaker lie to us?

So when Fox introduces us to Louis Meeks of Pavillion, Wyoming, a Wilford Brimley lookalike, a Vietnam vet, and all-around good egg, we're ready to believe his story of water pollution from nearby fracking. Just to be sure, Fox records Meeks and his friends talking about the value they place on honesty – "I mean, their word ain't no good!" Meeks said of the gas companies. "We's all raised that way. If your word ain't no good, you ain't no good," he said. It was all Fox could do to not break out his banjo and join in the folksiness. Who could possibly doubt such a pillar of honesty and American virtues – other than a predatory gas company, and maybe Dick Cheney?

To prove how the fracking companies poisoned his water, Meeks fills up an outdoor horse trough from a hose, scoops some out for Fox to smell. Fox says the smell of it instantly made him sick. Except that methane does not have a smell – that rotten eggs scent is added by natural gas companies as a warning in case of a gas leak in a home. Just in case that wasn't theatrical enough, Meeks holds a blow-torch to it. Fox's narration fills the time by listing unpro-nounceable chemicals. It's obvious: this honest cowboy survived Vietnam, only to be poisoned on his own farm. And if you had any lingering doubts, Meeks tells Fox that when he tried to drill a new water well – to get away from all the fracking poisons in that tap water – his new well exploded in a noisy geyser of natural gas, blasting millions of cubic feet of methane in the air, until a gas company came over to cap it for him. Fox got the video of it. Case closed.

It's not even the facts of the case that stay with you – Fox interviews a half dozen such people in the film, with different names, in different towns. The details aren't important. What's important is the emotional impact of talking to severely normal people, seeing them in their backyards, and then hearing their heartfelt accusations against the lying gas companies. The theatrical stunts – holding a blowtorch to a water trough – are just the icing on the cake. *Gasland* is not a fracking documentary. It's anti-fracking pornography.

It is a fact that when Meeks was drilling a water well, he struck a pocket of natural gas. But that had nothing to do with fracking – Meeks's home sits atop vast amounts of natural gas, in conventional underground reservoirs, not trapped in tiny pores of shale rock. A conventional gas well was drilled there in 1980, and Meeks's water well happened to drill into it – a rare occurrence, but predictable given that Meeks had a permit that allowed him to drill for water only 300 feet deep, and he decided to drill nearly double that depth, striking a gas pocket at 540 feet deep.

But what does any of that have to do with fracking – the closest frack job in the neighbourhood being 1,700 feet below the surface, separated by a quarter mile of rock?

Meeks says his water is contaminated, and Fox even sniffs it on camera and mentions some dangerous-sounding chemicals – case closed. But did any independent scientists test the water? Fox didn't say – for a good reason. Meeks's water was tested, repeatedly. Here's what the left-of-centre news agency ProPublica wrote when it sent a reporter to follow up with Meeks a year after *Gasland* aired: "By the standard commonly used to decide if water is safe to drink – the sort of test a homebuyer would require before signing a mortgage – Meeks's water was fine. It didn't contain heavy metals or arsenic or any of the handful of obvious contaminants that drinking-water specialists look for."[26]

ProPublica is no friend of fracking; it was founded and funded with a deliberate environmentalist and pro-regulation editorial position. Its reporter interviewed the Department of Environmental Quality (DEQ) and spoke with Mark Thiesse, the supervisor for water quality in the district. He tested Meeks's water five times. "We have not found hydrocarbons. We have not found fracking chemicals. We have found nothing out of the ordinary," he said.

Another left-wing magazine called *Counterpunch* – it describes its work as "muckraking with a radical attitude"

– followed up on *Gasland*'s spectacular story of the Meeks family too. It subtitled its report "Poisoning the Wells," so you can see where it's coming from. But it too reported no evidence of any fracking contamination of water. In addition to the government's water tests, *Counterpunch* says "the DEQ asked EnCana to test the Meeks water repeatedly, at least eleven times in between 2004 and 2007. The results always came back clean. The governor's office finally asked Meeks not to call any more."[27] It's true, as Fox showed, that EnCana had trucked in fresh drinking water to Meeks's home for a period of time, as a gesture of goodwill. ProPublica reported that EnCana spent $170,000 trying to help Meeks with his water worries, and he negotiated a confidential payment from them too. But that was the company's attempt to assuage a cranky neighbour – it was never the result of a failed water quality test, or any environmental order. It seems that Meeks agreed – in 2009, he reconnected the water well that he claimed was contaminated and began using it again in his house. Fox didn't mention that part either.

Donna Meeks, Louis's wife, might have given away the game when she told *Counterpunch* that she and her husband didn't even notice anything different about their tap water. She said they decided to become concerned when some of her co-workers no longer drank her coffee. She used to have an unusual practice of bringing water from her home "to town for the school office coffee pot," the reporter wrote. "Neither she nor her husband Louis noticed anything wrong until her co-workers stopped drinking the coffee." If Donna Meeks told Josh Fox the same story, he edited it out – it wouldn't fit the narrative of poisoned, stinking, flammable water if the only clue the Meekses had that they should be worried was because Donna's co-workers didn't drink her coffee any more.

Fox's movie was released in 2010. By then, a dozen or more tests had concluded that Meeks's water might not have been delicious to taste or smell, but there was no proof of fracking contamination in it. But the entertainment value of the cowboy with the blowtorch was just too good to muddle with any contrary facts.

Fox tapped into a different set of emotions for his interview with Lisa Bracken: sorrow, and the unexpected loss of her father to cancer.

Like Louis Meeks, her property has natural gas under it. Not from tiny pores in the rock, like frackable shale gas has, and not even from underground rock caverns of natural gas, like the Meekses' property. Bracken's property has what scientists call biogenic natural gas – created from rotting plants. "Biogenic methane is created by the decomposition of organic material through fermentation, as is commonly seen in wetlands," explains the Colorado Oil and Gas Conservation Commission.[28] It's not from deep down underground, inside rocks. It's from wet plant matter on the surface. Like a compost heap.

When Josh Fox met Lisa Bracken in 2009, she was seized with a theory that a fracked well had caused a "seep" of natural gas out of the ground, killing birds, rabbits, and other small creatures. So certain was she of the culprit that she collected these dead animals and stored them in her freezer until someone – anyone – could prove her theory that fracking had killed the animals.

Fox couldn't resist and asked to see Bracken's homemade morgue, and she happily complied, unwrapping the bagged, frozen-solid animals for his camera, all the while telling him of her dark nightmares. It's odd – but Bracken is so hyperkinetic and entertaining, she comes across as more quirky than anything.

Until she mentions that it's not just animals who died. Her father did too – and she blames that on fracked gas, too. Her theory: her father drank water from a creek that was bubbling with natural gas, and two years later he died of cancer of the pancreas.

All of a sudden the manic collection of animal corpses changes from quirky to terribly sad, like a grieving ritual not just for the lost animals, but for a lost father. Fox milks it for every ounce of pain – and then goes and stands by the side of the creek, alone, in his own mourning ritual.

It's shocking and enraging – and that's precisely the desired effect. But it's also completely untrue. No one knows why Bracken's father died of pancreatic cancer. The National Cancer Institute lists medical risk factors like smoking, drinking, diabetes, and genetic factors[29] – drinking water through which natural gas bubbled isn't among them. But science can't stack up against a daughter's grief.

But what Fox does with that grief can: he packages it and labels it, and blames it on fracking. And, like with the Meekses, you'd need a heart of stone not to hate EnCana a little bit after Bracken's story.

Unless you read the detailed reports commissioned by Colorado's oil and gas regulator, responding to Bracken's complaints. They took her complaints extremely seriously, with multiple visits, taking nine samples of soil, two ground-water samples, two surface water samples, and two samples of the natural gas, plants, and even rocks.[30] They performed a number of high-tech tests, even checking the isotopes of the gas – the "fingerprint" that each sample of gas has. Their conclusion: the gas on Bracken's property was natu-rally occurring, from decaying plants. Just as it had been the last time they visited.

The regulator visited her in 2004, 2007, 2009, and 2010. Same thing every time. She kept e-mailing;[31] they kept replying. Bracken claimed the water had a sheen on it, like a film of oil or gas – that's certainly what she told the media. But according to the lab test, it was the presence of "large quantities" of bacteria. It was water taken from stagnant beaver ponds. "These naturally occurring nuisance bacteria produce iridescent sheens on stagnant water, iron hydroxide (orange/rust colored) stained slime, black slime, and red, orange, and/or black particles," Debbie Baldwin wrote to Bracken in June of 2008 – a year before she met Josh Fox. Baldwin is the environmental manager for the state's Oil and Gas Conservation Commission. She had told Bracken this before; she had been there before; she had talked about swamps and biogenic gas before: "In 2004, during my inspections along West Divide Creek I observed this kind of biological activity at numerous locations and I believe that I pointed out occurrences to you at a couple of locations on your property. From the photographs you have provided us, it appears that what you are observing are deposits related to biological activity of these nuisance bacteria," she wrote. But for a grieving daughter, nursing a conspiracy theory about what killed her father, the answer couldn't be merely "cancer." It had to be cancer caused by fracking, by an evil company. The facts didn't support that theory. But Josh Fox did. He enabled Bracken's self-delusion. He was thrilled by the spectacle of her animal morgue. She was a political bullet that he wanted to shoot through the gun of his movie.

Not all of Fox's characters were as pitiful as Bracken. Some were motivated by other emotions – like the desire for fame. Or money.

There's a lot of money involved with fracking. Money for those who sign leases with gas companies, for the right to drill their land. But suing natural gas companies for millions of dollars – often for claims of environmental degradation – are another way to get paid. It's how EnCana finally got Louis Meeks of Wyoming to cork it, even though lab tests showed his water was fine.

And so, when Josh Fox visited the town of Dimock, Pennsylvania, he came into a place where dozens of local residents were suing the gas company there, called Cabot Oil and Gas. If you were a plaintiff in a multi-million-dollar lawsuit against a fracking company, you might just have an interest in using the media to put a little pressure on the defendant. So it wasn't hard for Fox to find plenty of people in that little town willing to swear on a stack of bibles that they'd been poisoned by Cabot. It was a perfect symbiosis.

Sometimes, the greed was easy to spot: one plaintiff suing Cabot put up lawn signs around their modest ranch house saying, "For Sale: $5,000,000."[32] For some, they thought a lawsuit would do for their humble lifestyle what a stray gunshot did for Jed Clampett in the TV show *The Beverly Hillbillies* – make them instant millionaires, instead of mere thousandaires as it would do for so many who signed a lease. It's tempting. But it's not reality: the local county property tax assessment of all properties of the eighteen litigants combined is just $2 million, for an average of just over $100,000 each.[33]

The Dimock plaintiffs were already media hounds before Fox visited them – even as he was taping in town, a TV news report came on about their claims of water poisoning. Fox wasn't the first media man the Dimock plaintiffs worked, but he was the most skilful. He didn't bother with boring things like actual lab reports that talk about parts per million. He went for the heartstrings – a mom who claimed all her four kids were sick and "couldn't handle eating anything for over a month," or another mom talking about how her cat kept vomiting, though Fox wasn't lucky enough to catch that on film. Cats are known to vomit from time to time, but Fox leaves the viewer with no possible other explanation besides fracking as the culprit.

The Dimock litigants were so persuasive – at least to the media, and media-sensitive politicians – that the Department of Environmental Protection (DEP) announced a plan to build a multi-million-dollar water pipeline into the town from miles away in Montrose, to allow the shutdown of all water wells in town. That staggeringly hasty decision certainly added pressure on Cabot to settle, but it enraged those in town who said their water was just fine, and they didn't want to be saddled with massive tax increases to pay for a solution to a non-problem.

Fox spoke with five different Dimock families – in a town of 1,500 – who all said the same things. Pretty persuasive; they couldn't all be hucksters, could they? Just to be sure that viewers wouldn't answer that question in the affirmative, Fox neglected to mention their lawsuits against Cabot or the massive gap between what their property was worth and what they were suing for.

Things were going pretty well for the anti-fracking activists in town, at least with regard to their media campaign. But then the EPA's water quality study was released, in May of 2012. It would have been wonderful news for those truly concerned about their health. But for those looking to sell their country cottage for $5 million, it was a disaster: the EPA declared the water to be just fine.

"This set of sampling did not show levels of contaminants that would give EPA reason to take further action," the EPA's Roy Seneca announced.[34] It had studied sixty-one wells – including those of the vomiting cat and the mom whose family couldn't eat for a month.

One of the Dimock water wells that was tested belonged to Craig and Julia Sautner, two of the real stars in Fox's movie. The Sautners gave themselves a nickname: "the Brown Juggers."[35] That's because they'd carry around jugs of brown water that they claimed came from their taps. The

Sautners had a travelling road show, campaigning against fracking not just in their hometown, but wherever they'd get speaking invitations. It became their job, like travelling minstrels. Along with their brown jugs, they'd reel off chemicals they claimed were in the water, put there by fracking – including "weapons grade uranium."

That travelling show, and all the press that covered it, and Josh Fox's movie, was ammunition in the Sautners' lawsuit against Cabot Oil and Gas. Their property was assessed at $132,000[36] – surely nuclear waste would be worth an extra zero on a cheque?

Filmmaker Phelim McAleer, whose film *Frack Nation* is a passionate rebuttal to *Gasland*, visited the Sautners right before the EPA water studies were released. He got the full razzle-dazzle from the Brown Juggers, though Craig Sautner seemed a bit confused when McAleer asked him if the jugs were just props, and if he could take a glass of water from their taps right there.

By chance, McAleer and his film crew were in Dimock right when the EPA all-clear was announced. In fact, he was standing at the side of the road when Julie Sautner happened to drive by and recognize him as the skeptic with the tough questions. She stopped to give him a warning – clearly unaware that McAleer's crew were in a vehicle not far away, filming their entire exchange.

"I'd like to tell you, if you put Craig in the movie, you will be sued," she said. McAleer asked Julie about the EPA report; she just kept denying it. It was a ridiculous moment: Sautner could have disagreed with the EPA's findings, but to outright deny that's what they found was absurd. So after a few rounds of that, she tried a different tactic: she railed against him personally, calling him a "turncoat" and saying he "turned against his own country." McAleer didn't rise to that bait, so Sautner took another tack: "I hope you make lots of money because you're going to need it. What do you think, we're small-time? I will have the NRDC and everybody else on you . . . I guarantee I come down on you like a ton of bricks." (The National Resources Defense Council, or NRDC, is an anti-fracking environmentalist lobby firm, with a budget of over $100 million and 350 lawyers on staff.[37])

At that moment, Julie Sautner noticed the cameras, pointed to them, and said, "You put me in there, so help me God it will be the last thing you do."

None of that worked. McAleer's crew didn't turn off the camera. So Sautner announced she was calling 911. That failed as well, at which point Sautner declared, "I am armed, I will tell you that," and began rummaging through her purse. That was enough to make McAleer take a step back and put his hands up, but not to stop talking, or for his camera crew to stop filming.

McAleer wasn't the only person who thought filming an encounter with the Sautners was wise. So did the EPA officials, when they met with the Sautners to formally bring them the news of their water tests and to talk with them about the results. Through a freedom of information request to the EPA, McAleer got a copy of that videotape, and it was a performance fitting for the closing night of the Sautners' drama. The EPA video shows Julie standing up and stomping out of the room, while Craig did his "just folks" routine one last time. "This is bull-crap, I'm sick and tired of this!" he says, as he thumps sheafs of paper on a table. "What happened to you people? You guys aren't the same as you were two months ago," he says.

In fairness, that's probably true; EPA officials are generally sympathetic to people who claim they've been poisoned by oil and gas companies. But once such claims have been proved hoaxes, a bona fide environmentalist would likely feel used – and concerned that noisy false claims and hoaxes could discredit other, genuine victims of pollution.

"Do you think I made this stuff up?" asked Craig. The EPA staff were too polite to answer, but began to pack their briefcases and go.

Soon, Cabot announced that most of the litigants settled out of court. As is usually the case, the dollar amount was confidential, but the company's CEO said the total value was "not a material item" on their balance sheet. Translation: it was a small, nuisance payment.

John Hanger, the former DEP secretary who bungled Dimock in the first place – the one who ordered the multi-million-dollar water pipe – is now a long-shot candidate for governor of Pennsylvania. His panicky reaction to complainers like the Sautners is what fuelled the fearmongers like Josh Fox. Now that the reality has finally caught up, Hanger sounds almost depressed about Dimock – though he certainly doesn't seem to take any blame for himself.

"The reputation of the state has been badly damaged at some level by the media coverage, some of it irresponsible

and some of it accurate. The state has suffered, the gas industry has suffered." You don't say? "There's certainly money to
be made from fighting, whether it's lawyers, consultants,
or fund-raising appeals. There's probably more to be made
out of a good old fight than a peaceful resolution."

Hanger was one of those profiteers; though he didn't
pocket any money personally, he certainly profited from the
media coverage of taking on a big, dirty gas company and
forcing it to cough up money for water supplies while the
matter was investigated.

Funny enough, Hanger's campaign website doesn't
mention Dimock at all; his policy statement on shale-gas
fracking does mention more enforcement bureaucrats and
regulations he created.[38] But the man most directly responsible for fracking's black eye – and for slandering the town
of Dimock – has looked at the weathervane and has concluded that most people in the state seem to like fracking,
very much. "Increased gas production made possible by the
safe development of the Marcellus shale gas has produced
thousands of jobs and saved natural gas consumers about
$1000 per year compared to 2008 prices. It also reduced
electricity prices significantly for Pennsylvania's families
and businesses." That doesn't sound like the John Hanger
who helped enforce a ban on drilling in Dimock.

Dimock was a critical building block in the propaganda
campaign against fracking, because the overreactions by
local officials were an authoritative confirmation that the
wildest accusations against the technology had some
merit, according to responsible authorities. It's no use that

bureaucrats like Hanger are full of remorse, and that the largest citizens' group isn't the cranky eighteen who sued, but the much larger "Dimock Proud" campaign.[39] It's good that its website now comes up in the top ten Google results for the town's name – though five others contain the debunked reports of water contamination. The town has served its purpose, and the media have, for the most part, moved on.

But where the Dimock story gave fracking fears a sense of substance, nothing compared to the sizzle of Fox's visit to Weld County, Colorado, where he met Mike Markham and Marsha Mendenhall. Fox's visit there is in all the film trailers for *Gasland* and the movie poster itself: because Markham could light his tap water on fire.

There's no way to minimize how shocking it is to turn on a tap, light a match, and see the water explode. It's startling to find fire where it's not supposed to be; it has a sense of magic, or the unnatural, to it, given that water normally quenches fire. And there's the pure shock value of it, as Markham scorches the hair on his arm. It's the one moment in the film when Fox's well-rehearsed sleepy persona really works: being calm and wry after someone just exploded their tap water is exactly the look that a zen hipster like Fox wants to achieve.

When you've got so much shock and awe in a scene, there's no point in bringing in pointy-headed scientists. Boring statistics or chemical equations don't do well on film in the best of times, but why go for science when you have pyrotechnics?

Except that, however startling it sounds, the phenomenon of burning tap water has been around far longer than fracking.

Burning Spring, New York, was named for the natural gas that bubbled up in the spring water, and could be lit, *Gasland* style. It was first mapped by French explorer Robert de La Salle in 1669, who was brought there by local Indians.[40] That's the first town with such a name in the United States, but hardly the only one. There's a Burning Springs, West Virginia, and a Burning Spring, Kentucky. You wouldn't think that Niagara Falls would need any additional tourist attractions, but an early attraction was a natural gas vent along the river that was lit for tourists, for a price.[41] European missionaries record Indian tribes showing them the phenomenon along Lake Erie as early as 1626.

Searching any newspaper database for "tap water on fire" or similar phrases finds no shortage of stories going back decades, and for long-standing newspapers like the *New York Times*, contemporary accounts as far back as the nineteenth century.[42]

Even George Washington once experimented with setting fire to the methane-bubbling water of New Jersey's Millstone River.[43] And Benjamin Franklin once wrote, "On the Inflammability of the Surface of certain Rivers in America," after observing the curious phenomenon himself.[44]

But if flaming faucets aren't caused by fracking, what causes them?

Naturally occurring gas, called "biogenic" gas, from decaying plant matter. After *Gasland* debuted, the Colorado

Oil and Gas Conservation Commission (COGCC) put out a four-page technical document explaining why Markham's tap water was burning.[45] It explained that Markham's well, and those of two other local property owners who appeared in *Gasland*, had their water investigated in 2008 and 2009. It had nothing to do with fracking, but the fact that Markham was living on top of a lot of natural gas, and that the "water well completion report for Mr. Markham's well shows that it penetrated at least four different coal beds."

Geological surveys taken in 1976, 1983, and 2001 showed that the aquifer contains "troublesome amounts of . . . methane." Markham wasn't lying when he lit his water on fire. Fox wasn't exaggerating when he filmed it. It's just that, like all the towns named Burning Springs, it had nothing to do with fracking, or even regular oil and gas production. There just are some parts of the world where natural gas gurgles up through the land or the water.

The COGCC document is exhaustive and completely persuasive. For those who have seen it, and could plough through its four pages of technical jargon. But *Gasland* is a visual film that takes no scientific sophistication. The flaming faucet scene can be found everywhere on the Internet. Millions of people have seen it – almost none of whom will ever read the COGCC's painstaking technical explanation.

That doesn't mean you can't find people who will look at you – or probably the president – with a straight face and insist that their water tastes different, or that they can light it on fire. Some of them may well be fraudsters, ginning up evidence as Steven Lipsky did – connecting his garden house to a gas vent, with the help of an environmental activist (bearing a vendetta against a local gas company) to put on a show.[46] Maybe some think it will lead to a rich class-action lawsuit someday. Still others may have convinced themselves that it's true that shale exploration harmed their water, even if all the evidence shows that fracking fluid does not contaminate groundwater and, beyond that, that it is actually physically impossible for it to do so.

Chapter Thirteen

LUDDITES

Some people attack fracking out of genuine fear for their health and safety. They're good-faith critics, often political amateurs. Better funded and better organized are the bad-faith critics, motivated by their own competitive interests – like Gazprom, Qatar, and Iran. They're often paid professionals.

But still others are motivated by ideology, a personal philosophy, almost a religion, that believes mankind should not be so technologically advanced. It's not natural or normal, and Mother Earth will strike back at us for living too large for too long.

To these people, fracking is the worst possible development: an almost miraculous technology that virtually creates something out of nothing, that sets back the main narrative of the environmentalist movement, namely that we're all going to run out of stuff, so we should start living on less.

Environmentalists preach "reduce, reuse, recycle" – but fracking makes reducing unnecessary, because it ends shortages of energy. It's not just that fracking is a change in technology. It's that fracking requires a change in philosophy. The moral condemnation that environmentalism has attached to material consumption doesn't make sense any more if we don't have to ration energy. If we're not going to run out of energy, we don't have to ration it for economic reasons. And if natural gas is the cleanest of the fossil fuels, we don't have to ration it for environmental reasons either.

This isn't the first time technological progress has challenged the moral order. While fracking promises an energy revolution, it's of a much smaller magnitude than the original Industrial Revolution itself.

And just as the fracking revolution has generated a backlash, so did the original Industrial Revolution. In the early 1800s, a group of unemployed hand knitters lashed out against machine-knitted clothing – and most other trappings of the new industrial age. They wanted the machines to be un-invented, especially the mechanical looms that could do the work of dozens of manual labourers. And if that couldn't be done – and of course, it couldn't – and if ordinary Englishmen wouldn't choose to buy expensive, hand-knitted socks, instead of affordable, machine-made socks, these knitters would smash the machines.

They'd riot. Not to loot and steal – but to destroy. To de-industrialize.

Their hero was a man named Ned Ludd. He was a weaver, who in a fit of passion destroyed two mechanical stocking frames. The story became the stuff of legend, and Ned Ludd became the symbol of an earlier, simpler era, raging against the machine. That's where we get the term "Luddite" from, a word we use as a joke to describe people who are slow adopters of modern technology, like someone who doesn't like mobile phones.

But the real Luddites didn't just reject technology. They tried to stop it, vainly trying to turn back time and progress, for reasons of personal economic interests, but just as much for psychological reasons too. In the early 1800s, Luddite riots were so widespread, they became as significant a threat to public order in Great Britain as Napoleon himself, if you measure by the number of soldiers required to put down the riots.

In 1817, three hundred armed Luddites set out to smash machines and protest against an economic recession. Their leader, Jeremiah Brandreth, was caught and tried for treason and hanged. And in a barbaric flourish, after he was executed, his head was cut off with an axe. That's how serious the Luddite riots were, and how brutally they were suppressed. But the economic theories, the de-industrialization dream, and the emotional fear of technological progress still live on.

To be sure that's not the philosophy of most environmentalists today. But it is, for some, the logical conclusion of an increasingly anti-industry belief system.

Today you might hear them calling themselves "de-growth" activists. They're not necessarily religious, but they espouse an apocalyptic vision of the world where technology is leading to ruin. Some call their belief "deep green" because even the regular green types – the sort that lobby for wind power and solar power – are dismissed as unserious, or men of half-measures.

Deep green activists don't think there is any way we can be richer and healthier and happier, even with eco-friendly technology. They think that's fooling ourselves, and the sooner we come to accept it, the sooner we can start de-industrializing our economies and "transitioning" back to the kind of life people led before industrialization changed the world.

Deep greens believe we've hit a "peak." Peak-oil theory was popular about a decade ago, when oil prices were rapidly rising and environmentalists, including Al Gore, insisted we were at, or near, the limits of our ability to keep finding new oil. Nobody talks about peak oil much any more. Not since technology – including fracking – has given the world more oil proven reserves than at any time in human history. In fact, in the summer of 2013, the Oil Drum, a website that acted as a central clearing house for news and commentary on peak oil, announced it would cease activity after eight years due to "scarcity of new content caused by a dwindling number of contributors." Even the one-time peak-oil true believers can't rouse themselves to make their phony case any more.

But degrowthers don't just talk about peak oil: they talk about peak "everything." They say we've hit a wall: that the earth can't sustain any more humanity. We're at maximum food production, maximum energy production, maximum mineral usage, maximum urban sprawl, maximum everything. To degrowthers, humans aren't the ingenious solvers of problems – we are the problem. We are a plague on the environment. And we need to be stopped.

Of course, we've heard all this before: Thomas Malthus famously predicted that humanity had reached "peak everything" in the eighteenth century. That's when there were 900 million people on the planet. Stanford University's Paul Ehrlich predicted it, in his book *The Population Bomb*, back in 1968. In 1968, the world's population was 3.5 billion. Today, there are more than seven billion in the world and, overall, they are better fed, better clothed, better housed, and richer, than when either Malthus or Ehrlich predicted the whole thing would collapse in on itself.

But all these philosophies live on. It's as if some people feel guilty for the fact that humanity is doing so much better; that there must be some great retribution looming for the fact that we've not just persevered, but prospered. That's why they hate new technological discoveries that unleash yet more prosperity and potential: discoveries like fracking. Fracking doesn't just powerfully undermine everything they believe in, it enriches us and helps us even more, which must mean that we'll pay an even dearer price when that great cosmic Judgement Day comes to make us pay for thriving.

The degrowthers don't just talk about this stuff: they've actually begun to prepare for the great calamity. In the United Kingdom, the land of Ned Ludd, they have begun christening "transition towns" – places they're working to actively de-industrialize. Places where people can begin living like the subsistence-economy ascetics they're so sure we'll all end up like.

In Totnes Transition Town, they're planting nut trees as their "emergency" food source for when the world's agriculture collapses. It's the fundamentalist version of the locavore trend and the 100-mile diet fad. Add in preparations for a post–fossil fuel future, and you've got a place that starts to sound a bit like the North Korean economy.

The de-growthers are wrong. Technology plus ingenuity plus freedom makes the world better – always has. But their cultish vision has always been deeply pessimistic. They hate fracking because fracking is an energy solution that will make us healthier, richer, and happier. Well, most of us will be happier. There will likely be Luddites among us four hundred years from now, too.

S o which future will it be? The future promised by fracking's advocates, where a shale-gas revolution will provide the world with cheap, plentiful, clean energy, freeing consumers – and even entire countries – from OPEC-style cartels? Or will it be the future promised by fracking's opponents, where "unconventional" oil and gas is kept locked in the ground out of an abundance of caution, and energy is either bought dearly from Russia, Iran, and Qatar, or produced expensively, by highly subsidized wind turbines and solar panels?

The answer is: both.

The fracking genie can't be put back in the bottle in North America. Approximately 90 per cent[1] of all oil and gas wells in the United States are fracked. The modern process was invented and perfected in the United States. It may sound newfangled or dangerous in media bubbles like New York and Los Angeles, but in U.S. states where oil and gas is produced, fracking can hardly even be called unconventional any more. Pockets of political resistance will likely continue – that's the beauty of fifty U.S. states, each with their own laws – but nationally, it's economically and politically entrenched, and one of the only things powering the United States out of its recession.[2] Even those U.S. states that continue to resist fracking their own shale gas will still benefit from the cheaper gas piped in from other states. And unemployed young people will continue to cross state lines, making their homes in fracking boom towns. Four of the five fastest-growing cities in the United States are located in shale-gas bonanzas[3] – two in Texas, and two in North Dakota.

It is the same thing in Canada: in the three western provinces, fracking is mainstream, and plans to export LNG to Asia got a staggering boost with the announcement of a $35-billion[4] export facility in British Columbia – the largest single capital investment in Canadian history. As with the United States, anti-fracking resistance is greatest where familiarity with it is the lowest. But even if the anti-fracking riots in eastern provinces like New Brunswick are ultimately successful, and if the "temporary" moratoriums in Nova Scotia, Newfoundland, and Quebec become permanent, those provinces will still benefit from using cheap shale gas from elsewhere in North America. And young workers in Atlantic Canada will still seek out high-paying jobs in the industry – they'll just have to move to western Canada to do it.

Fracking won't be stopped in North America. But much of the anti-fracking propaganda movement in America is just political theatre for the benefit of Europeans – that's why Al Jazeera sent a team to tiny Rexton, New Brunswick, to cover the anti-fracking violence there. Qatargas can continue to be the biggest LNG exporter, and Gazprom can continue to be the largest natural gas company in the world – with staggering profits – just fine without shipping to the U.S. market. But if the United States starts elbowing in to former Warsaw Pact satellite states like Poland and Ukraine, and if North American surpluses start breaking in to the Japanese and Korean markets, that's where the problems begin. That's the market "gas OPEC" wants to protect.

So which will it be?

The United Kingdom seems determined to embrace fracking, despite a passionate and well-funded opposition. This is in part a reaction to the costly experiment of the former Labour government's massive subsidies of wind power, and the public backlash over staggeringly high energy costs, particularly in wintertime cold snaps. But there's also a historical reason; the ties between Britain and America are deep, covering politics and commerce and geopolitics. If Gazprom is against something and America is for it, there's a predisposition for Britain, a keystone member of NATO, to side with its best ally. And the historical fact that the United Kingdom's own Industrial Revolution came about because the island country happened to be sitting on an underground mountain of coal – well, that's a cultural memory of energy self-sufficiency that may be an added psychological boost for finding British energy, as opposed to importing it. It's just not the British way to be forced to depend on the goodwill of Vladimir Putin, or the Emir of Qatar, for something as strategic as energy.

But what about Eastern Europe? What about the countries specifically targeted by the U.S. State Department's fracking technology transfer program, as an instrument of anti-Putin foreign policy?

Here geology plays an important role. Three major North America–based developers – ExxonMobil, Marathon Oil, and Talisman Energy[5] – have abandoned their exploration for shale gas in Poland after challenging tests. And some geological surveys say the amount of frackable natural gas is less than promised.[6] But it remains a high priority for Poland's Western-leaning government, desperate to avoid falling back into Russia's orbit. It's a popular position in the country, where only 3 per cent of the public say they're strongly opposed to fracking, compared to majorities in some Western European countries. At the very least, Poland's vocal interest in fracking could lead to significant discounts from Gazprom when the current contract with Poland expires in 2019; at most it could render those contracts unnecessary.

That's certainly Russia's fear – and Poland's counter-intelligence agency, the ABW, has warned that Russian agents are engaged in widespread spying on Polish shale gas projects. They know what Poland knows: if fracking ever caught on in Poland – if it were to be an economic success, all the while avoiding the horror-story environmental scenarios promised by hysterical opponents – it would become a political success, too. It would be hard for other Gazprom-dependent European countries to defend their continued dependence on Russia if Poland can break free.

Russia's Kremlin-backed English-language news agency, RT, openly frets about this potential domino effect – lavishing praise on anti-fracking activists in the West, just like Qatar's Al Jazeera does.[7]

In other words, it's like the Cold War all over again – spies; propaganda; spheres of influence. But it's not military dominance now. It's economic and political dominance. Will Eastern Europe once more be welded to Moscow and its increasingly authoritarian rule? Or will it join the West, increasing its economic integration with the West, and avoiding the risk of pricing blackmail, or even sudden pipeline blockades?

That's the front line of this war, really: Ukraine and Poland, two of the most invaded, abused countries in the world, with educated, Western-oriented people who have too often been subjugated by eastern empires.

Sure, the political battles in places like France are interesting, but there's no real risk of France falling under Vladimir Putin's thumb – banning fracking is just one of a hundred anti-industry rules and regulations in that socialist republic. But to Ukraine and Poland, it's about sovereignty itself.

France is a G8 economy with a per-capita GDP of about $40,000 a year. Saddling each Frenchman with an extra few hundred bucks a year by importing Gazprom gas, instead of developing France's own shale gas, is the kind of luxurious indulgence that a rich country can afford. Not so much for Poland, with a per-capita GDP of $13,000. Ukraine's economy is even poorer – less than $4,000 per person, per year. Finding an alternative to Gazprom's extortionate monopoly isn't a matter of political fashion, as it might be for Parisians. In Ukraine, it's a matter of the most basic standard of living.

So what will these countries on the front lines of the fracking war do? It's impossible to predict the future. But there is cause for optimism. Poland and Ukraine have had a long and terrible history of being invaded – they were too often the buffer between rival empires in Europe, and if they weren't being sacked and looted by one invader, they were by the other.

The most famous was the 1939 Nazi–Soviet pact. Publicly, that treaty was a non-aggression pact between two dictatorships. But there was a secret provision to carve up Poland between the two powers. Not much different from how the Russian czars carved up the country, or even Napoleon before them. Ukraine, too, was traded back and forth between tyrants. It's difficult to know who was more brutal toward the country: Stalin or Hitler.

But here's the difference: whereas Poland and Ukraine had little chance of defending their national independence against the Wehrmacht or the Red Army, the choice to make themselves economically independent is within their power.

Of course, Russian propaganda and spies are trying to undermine that choice. But that's far less of a threat than an actual invasion by Russia. For the first time in centuries, Poland and Ukraine truly are the masters of their fate – or could be.

The geology of their countries say energy independence through fracking is possible. The economics of fracking – not just the lower cost of gas, but all the jobs that come with it – make it attractive, attractive enough that powerful, experienced businesses from America are willing to give it a try.

The politics may be challenging. But the political payoff could be enormous. The great heroes of Poland and Ukraine are invariably those who have driven out foreign meddlers, to give the countries back their sovereignty. In the case of Jan Sobieski, the Polish king and military commander, he didn't just save Poland, but all of Europe, by smashing the Ottoman siege of Vienna and driving the Turks out of Europe.

Not even a modern Jan Sobieski could smash Vladimir Putin's nuclear-tipped military. But this time no military victory is needed – just the political will to drill for gas.

That's the paradox of independence: it can't really be bestowed upon a country. Sovereignty is something that a nation can't be given – it has to be taken. No U.S. State Department policy, no U.S.-based energy company boardroom can tell Ukraine and Poland to become independent. It's a choice they have to make.

After two generations of a Cold War, Western Europe kept its promise, through NATO, of one day helping to liberate eastern Europe from Russia's steely grip. It would be historically poetic if, by bringing fracking to Europe, Ukraine and Poland returned the favour, by joining the shale-gas revolution that will ultimately free much of the world from economic domination.

Acknowledgements

Let me acknowledge the faith that Doug Pepper of McClelland & Stewart has in me. Thanks to Jenny Bradshaw for her meticulous and patient editing of the book, and to Josh Glover for spreading the word about it. And a big thank you to Kevin Libin for brainstorming to fact-checking and everything in between.

Introduction

1. The United States consumes approximately 25 trillion cubic feet per year: http://www.eia.gov/dnav/ng/hist/ n9140us2a.htm.
2. http://www.spe.org/jpt/print/archives/2010/12/10Hydraulic. pdf.
3. http://www.capp.ca/getdoc.aspx?DocId=210903&DT=NTV.
4. http://www.eia.gov/todayinenergy/detail.cfm?id=2170.
5. http://www.eia.gov/todayinenergy/detail.cfm?id=3750.
6. http://www.eia.gove/todayinenergy/detail.cfm?id=7550.
7. http://www.eia.gov/dnav/pet/pet_move_impcus_a2_nus_ epco_imo_mbblpd_m.htm.

Chapter 1: Plentiful, Cheap, Clean, Democratic Energy

1. http://www.halliburton.com/public/projects/pubsdata/ Hydraulic_Fracturing/fracturing_101.html.
2. https://www.asme.org/engineering-topics/articles/fossil-power/fracking-a-look-back.
3. U.S. Energy Information Administration, "Global natural gas prices vary considerably," Sept. 30, 2011, chart: "Trends in natural gas spot prices at major global markets," http://www.eia.gov/todayinenergy/detail. cfm?id=3310.
4. Nataliya Vasilyeva, AP via *The Guardian*, "AP Enterprise: Russia oil spills wreak devastation," http://www.guardian. co.uk/world/feedarticle/10001530.
5. http://www.iea.org/co2highlights/co2highlights.pdf, p. 57. Energy Information Administration (http://www.eia.gov/ environment/emissions/carbon/).

Chapter 2: An OPEC for Gas

1. U.S. Department of Energy, https://www1.eere.energy. gov/vehiclesandfuels/facts/2013_fotw781.html.
2. Ibid.
3. Terry Macalister, "Russia, Iran and Qatar announce cartel that will control 60% of world's gas supplies," *The Guardian*, October 22, 2008, http://www.theguardian. com/business/2008/oct/22/gas-russia-gazprom-iran-qatar.
4. Center for Energy Economics, "Brief History of LNG,"

http://www.beg.utexas.edu/energyecon/lng/LNG_
introduction_06.php.

5. David Black, "LNG ships in big demand – for now," *The National*, April 11, 2012, http://www.thenational.ae/
business/industry-insights/shipping/lng-ships-in-big-
demand-for-now.

6. "Iran eyes $4bn gas pipeline to Europe," Upstream
Online, September 24, 2008, http://www.upstreamonline.
com/live/article1163492.ece.

7. http://www.bloomberg.com/news/2011-11-25/oil-tanker-
glut-means-number-valued-as-scrap-jumps-fivefold-to-
101-in-year.html.

8. http://www.tankersinternational.com/fleet_list.
php?type=1&order=year_down.

9. http://energyforumonline.com/764/how-many-oil-
tankers-operate-in-the-world/.

10. http://unctad.org/en/PublicationChapters/Chapter%202.
pdf.

11. http://www.economist.com/node/21558456.

12. http://shipbuildinghistory.com/today/highvalueships/
lngfleet.htm.

13. http://unctad.org/en/PublicationsLibrary/rmt2012_en.pdf.

14. http://www.ferc.gov/market-oversight/othr-mkts/lng/othr-
lng-wld-pr-est.pdf.

15. http://www.bbc.co.uk/news/world-asia-24099022.

16. http://blogs.platts.com/2012/12/04/japan_lng/.

17. http://www.businessweek.com/news/2013-01-22/japan-may-
save-30-percent-on-power-by-restarting-reactors-ieej-says.

18. http://www.bbc.co.uk/news/world-asia-23177536.

19. http://ferc.gov/industries/gas/indus-act/lng/lng-existing.pdf.

20. http://www.kitimatlngfacility.com/Project/project_
description.aspx.

21. http://www.ferc.gov/market-oversight/mkt-gas/overview/
ngas-ovr-lng-wld-pr-est.pdf.

22. Shipbuilding History, updated April 23, 2013, http://
shipbuildinghistory.com/today/highvalueships/
lngactivefleet.htm.

23. Dan Sabbagh, "Al-Jazeera's political independence ques-
tioned amid Qatar intervention," *The Guardian*, September
30, 2012, http://www.theguardian.com/media/2012/sep/30/
al-jazeera-independence-questioned-qatar.

24. Nicholas Noe and Walid Raad, "Al-Jazeera gets rap as
Qatar mouthpiece," Bloomberg, April 9, 2012; "Dozens
resign from Al Jazeera citing pro-Brotherhood bias, editorial control from Qatar," *The Tower*, July 8, 2013, http://
www.thetower.org/dozens-resign-from-al-jazeera-citing-
pro-brotherhood-bias-editorial-control-from-qatar/.

25. AlJazeera.com.

26. International Energy Agency, "FAQs: Natural Gas," http://
www.iea.org/aboutus/faqs/gas/.

27. CIA World Factbook, https://www.cia.gov/library/
publications/the-world-factbook/rankorder/2004rank.html.

28. Gas Exporting Countries Forum, website: http://www.
gecf.org/aboutus.

29. http://www.eia.gov/finance/markets/supply-opec.cfm.

30. Roman Kupuchinsky, "Russia: Algeria deal revives talk of
gas cartel," Radio Free Europe Radio Liberty, August 14,
2006, http://www.rferl.org/content/article/1070561.html.

31. Parisa Hafezi, "Russia, Iran, Qatar agree to form 'big gas troika,'" *The Globe and Mail*, October 21, 2008, http://www.theglobeandmail.com/report-on-business/russia-iran-qatar-agree-to-form-big-gas-troika/article1064417/.

32. Terry Macalister, "Russia, Iran and Qatar announce cartel that will control 60% of world's gas supplies," *The Guardian*, October 22, 2008.

33. "Shmatko predicts strong role of GECF," RT.com, December 11, 2009, http://rt.com/business/shmatko-gecf-opec-gas/.

34. "Gas exporters defend producer-countries' interests," The Voice of Russia, December 17, 2009, http://voiceofrussia.com/2009/12/17/3072817/.

Chapter 3: Gazprom: How the Shale Gas Revolution Weakens Russia's Energy Monopoly

1. Russia Today, "Court ruling sidetracks Gazprom's new $226 million HQ in St Petersburg," June 18, 2013, http://rt.com/business/gazprom-headquarters-cultural-area-880/.

2. Ahmed Mehdi, "Putin's Gazprom problem," *Foreign Affairs*, May 6, 2012, http://www.foreignaffairs.com/articles/137615/ahmed-mehdi/putins-gazprom-problem.

3. Gazprom website, "Gazprom in Questions and Answers: Gazprom in Foreign Markets," http://eng.gazpromquestions.ru/?id=4.

4. Ibid.

5. Gazprom website, "About Gazprom: Fields," http://www.gazprom.com/about/production/projects/deposits/.

6. "Gazprom acquires Kyrgyzstan's gas network, pledges secure supply," Natural Gas Europe, July 29, 2013, http://www.naturalgaseurope.com/gazprom-takes-over-kyrgyzstans-gas.

7. Kostis Geropoulos, "Gazprom strengthens its grip on Turkey," New Europe Online, February 12, 2012, http://www.neurope.eu/article/gazprom-strengthens-its-grip-turkey.

8. Michael Lelyveld, "Russia's Gazprom misses gas target," Radio Free Asia, July 1, 2013, http://www.rfa.org/english/commentaries/energy_watch/gazprom-07012013102351.html.

9. Mehdi, "Putin's Gazprom problem."

10. Charles Recknagel, "Interview: U.S. economist Judy Shelton skeptical of customs union," Radio Free Europe / Radio Liberty, September 4, 2013, http://www.rferl.org/content/russia-armenia-customs-union-reaction-judy-shelton/25095385.html.

11. Stefan Schultz and Benjamin Bidder, "Under pressure: Once mighty Gazprom loses its clout," Spiegel Online, February 1, 2013, http://www.spiegel.de/international/world/gazprom-gas-giant-is-running-into-trouble-a-881024.html.

12. Ibid.

13. Dave Keating, "Commissioner urges EU to face down Russia on energy," European Voice, November 10, 2012, http://www.europeanvoice.com/article/imported/

commissioner-urges-eu-to-face-down-russia-on-energy/75339.aspx.

14. William V. Lamping, "20 years later, Eastern Europe struggles for energy independence from Russia," Europa Bezpieczeństwo Energia, September 25, 2012, http://ebe.org.pl/komentarz-tygodnia/20-years-later-eastern-europe-struggles-for-energy-independence-from-russia.html.

15. Ibid.

16. Ibid.

17. Philip Dewhurst, "Fuelling Gazprom's warmer image," International Public Relations Association, August 8, 2008, http://www.ipra.org/frontline/08/2008/fuelling-gazprom-s-warmer-image.

18. YouTube video available at http://www.youtube.com/watch?v=xGbI87tyr_4&list=FLNTjdQv5WHgPBkz-V7-1uMA&index=1.

19. Excerpts from a speech delivered by Alexander Medvedev, Director General of Gazprom Export, at the "Gas of Russia – 2012 Forum," Blue Fuel, December 2012, http://www.gazprom-mt.com/WhatWeSay/Lists/PublicationsList/17%20-%20Blue_Fuel%20newsletter%20-%204Q%20-%202012.pdf.

20. John Daly, "Green eyed Gazprom attacks Turkmenistan's natural gas resource," OilPrice.com, November 24, 2011, http://oilprice.com/Energy/Natural-Gas/Green-Eyed-Gazprom-Attacks-Turkmenistans-Natural-Gas-Resources.html.

21. Ibid.

22. Andrew Rettman, "Kremlin backs its dog in EU–Gazprom fight," EU Observer, July 9, 2012, http://euobserver.com/economic/117474.

23. Michael Kahn, Braden Reddall, and Gabriela Baczynska, "Insight: Poland's shale gas play takes on Russian power," Reuters, February 9, 2012, http://www.reuters.com/article/2012/02/09/us-poland-shalegas-idUS-TRE8180PM20120209.

24. Aviezer Tucker, "New cold war over shale gas," *The Washington Times*, July 13, 2012, http://www.washingtontimes.com/news/2012/jul/13/new-cold-war-over-shale-gas-russia-inflames-enviro/.

25. Benno Spencer and Brian Hansen, "Russia's Gazprom skeptical of US-led shale gas boom," Platts/McGraw Hill Financial, July 23, 2012, http://www.platts.com/latest-news/natural-gas/Washington/FEATURE-Russias-Gazprom-skeptical-of-US-led-shale-8545823.

26. Kevin Begos, "Next Cold War? Gas drilling boom rattles Russia," AP, September 30, 2012, http://bigstory.ap.org/article/next-cold-war-gas-drilling-boom-rattles-russia.

27. Tucker, "New cold war over shale gas."

28. http://www.youtube.com/watch?v=NDpopfFMci8.

29. James Herron, "Gazprom, the unlikely environmental evangelist," *The Wall Street Journal*, February 10, 2010, http://blogs.wsj.com/source/2010/02/10/gazprom-the-unlikely-environmental-evangelist/.

30. Jacob Gronholt-Pedersen, "Russia sounds alarm on shale gas," *The Wall Street Journal*, November 30, 2011, http://

blogs.wsj.com/emergingeurope/2011/11/30/russia-sounds-alarm-on-shale-gas/.

31. Jake Rudnitsky and Stephen Bierman, "Exxon fracking Siberia to help Putin maintain oil clout," Bloomberg News, June 14, 2012, http://www.businessweek.com/news/2012-06-13/exxon-fracking-siberia-to-help-putin-maintain-oil-clout.

Chapter 4: Criticisms of Fracking: Contaminated Groundwater

1. Josh Fox, "An open letter to President Obama from Gasland director Josh Fox," July 9, 2013, EcoWatch.com, http://ecowatch.com/2013/josh-fox-open-letter-president-obama/.
2. Dan Springer, "Energy in America: EPA rules force Shell to abandon oil drilling plans," April 25, 2011, Fox News, http://www.foxnews.com/us/2011/04/25/energy-america-oil-drilling-denial/.
3. Jennifer A. Dlouhy, "Houston's Noble Energy lands first post-ban offshore drilling permit," Houston Chronicle, February 28, 2011, http://www.chron.com/business/energy/article/Houston-s-Noble-Energy-lands-first-post-ban-1687958.php.
4. Video of Barack Obama's meeting with San Francisco Chronicle editorial board, January 17, 2008, http://www.youtube.com/watch?v=DpTIhyMa-Nw.
5. Juliet Eilperin, "Is Obama waging a 'war on coal'?" The Washington Post, June 25, 2013, http://www.washingtonpost.

com/blogs/the-fix/wp/2013/06/25/is-obama-waging-a-war-on-coal/.

6. Valerie Volcovici, "Report says Keystone pipeline would not up U.S. greenhouse gas emissions," Reuters, August 8, 2013, http://uk.reuters.com/article/2013/08/08/usa-keystone-carbon-idUKL1N0G926B20130808.

7. John M. Broder, "Report may ease path for new pipeline," *The New York Times*, March 1, 2013, http://www.nytimes.com/2013/03/02/us/us-report-sees-no-environmental-bar-to-keystone-pipeline.html.

8. Environmental Protection Agency, Memorandum from Lisa P. Jackson, Administrator, to All EPA Employees, January 12, 2010, http://blog.epa.gov/administrator/2010/01/12/seven-priorities-for-epas-future/.

9. Henry I. Miller, "The EPA's Lisa Jackson: The worst head of the worst regulatory agency, ever," Forbes.com, January 30, 2013, http://www.forbes.com/sites/henrymiller/2013/01/30/the-epas-lisa-jackson-the-worst-head-of-the-worst-regulatory-agency-ever/.

10. Jeff McMahon, "EPA chief resigns: Declared carbon dioxide a pollutant," Forbes.com, December 27, 2012, http://www.forbes.com/sites/jeffmcmahon/2012/12/27/epa-administrator-resigns-declared-carbon-dioxide-a-pollutant/.

11. "Top EPA official resigns after 'crucify' comment," Fox News, April 30, 2012, http://www.foxnews.com/politics/2012/04/30/top-epa-official-resigns-after-crucify-comment/.

12. EPA, "Evaluation of impacts to, underground sources of,

drinking water by hydraulic, fracturing of coalbed, methane reservoirs," June 2004, http://nepis.epa.gov/Adobe/PDF/P100A99N.PDF.

13. "Regulatory Statements on Hydraulic Fracturing, Submitted by the States," June 2009, http://www.iogcc.state.ok.us/Websites/iogcc/Images/2009StateRegulatory StatementsonHydraulic%20Fracturing.pdf.

14. Ground Water Protection Council, "State oil and gas agency groundwater investigations and their role in advancing, regulatory reforms," August 2011, http://fracfocus.org/sites/default/files/publications/state_oil__gas_agency_groundwater_investigations_optimized.pdf.

15. The Big Well Museum and Visitors Center, http://kansastravel.org/greensburgbigwell.htm.

16. Peter C. Glover, "Ten fracking things everyone should know," EnergyTribune.com, April 21, 2011, http://www.energytribune.com/7499/ten-fracking-things-everyone-should-know#sthash.Ie8ppEXO.dpbs.

17. FracFocus Chemical Disclosure Registry, Ground Water Protection Council and Interstate Oil and Gas Compact Commission, "Hydraulic fracturing: The process," http://fracfocus.org/hydraulic-fracturing-how-it-works/hydraulic-fracturing-process.

18. Ibid.

19. Letter from EPA Administrator Carol M. Browner to David A. Ludder, General Counsel, Legal Environmental Assistance Foundation, Inc. May, 5, 1995, http://energyindepth.org/docs/pdf/Browner-Letter-Full-Response.pdf.

20. The Mother's Project, Endorsements, http://www.
mothersforsustainableenergy.com/environmental-
threats/2012/07/19/calling-on-mothers-to-save-colorado-
from-hydrofrackingsign-on-now.
21. Claudio Bruffato et al., "From mud to cement – building
gas wells," *Oilfield Review* 15, 3 (Autumn 2003), 62–76.
22. Yoko Ono, Letter: "Concerns about the safety of drilling,"
The New York Times, December 25, 2012, http://www.
nytimes.com/2012/12/26/opinion/concerns-about-the-
safety-of-fracking.html?emc=eta1&_r=3&.

Chapter 5: Criticisms of Fracking: Using Too Much Water

1. Tom Kentworthy, "Fracking is already straining U.S.
water supplies," ThinkProgress.com, June 15, 2013, http://
thinkprogress.org/climate/2013/06/15/2163531/fracking-is-
already-straining-us-water-supplies/.
2. Douglas Main, "How fracking is drying up one unlucky
Texas town," TakePart.com, August 16, 2013, http://www.
takepart.com/article/2013/08/14/fracking-strains-water-
supplies-oil-or-water.
3. Siobhan Courtney, "Fracking: A dehydrated UK, watered
only by capitalism," Al Jazeera, May 7, 2012, http://www.
aljazeera.com/indepth/opinion/2012/05/2012514505415433.
html.
4. David Blackman, "Water for fracking, in context,"
Forbes.com, July 1, 2013, http://www.forbes.com/sites/
davidblackmon/2013/07/01/water-for-fracking-in-context/.

5. Rusty Todd, "Why the grass should not always be greener," *The Wall Street Journal*, June 28, 2013, http://online.wsj.com/article/SB100014241278873246375045785 68533026520790.html. The Texas Water Development Board says of 6.8 billion gallons used residentially in Fort Worth every year, 80 per cent to 90 per cent is used on lawns: 6.8 billion x 80% = 5.4 billion. 5.4 billion / 365 days = 14,794,520.

6. Erik Mielke, Laura Diaz Anadon, and Venkatesh Narayanamurti, "Water Consumption of Energy Resource Extraction, Processing and Conversion," Energy Technology Innovation Policy Research Group, Harvard Kennedy School, Belfer Center for Science and International Affairs, October 2010, http://belfercenter. ksg.harvard.edu/files/ETIP-DP-2010-15-final-4.pdf.

7. Joan F. Kenny et al., "Estimated Use of Water in the United States in 2005," United States Geological Survey Circular 1344, 2005, http://pubs.usgs.gov/circ/1344/.

8. U.S. Environmental Protection Agency, "How We Use Water in These United States," http://esa21.kennesaw. edu/activities/water-use/water-use-overview-epa.pdf.

9. U.S. Department of Energy, Quarterly Report, "Comprehensive Lifecycle Planning and Management System for Addressing Water Issues Associated with Shale Gas Development in New York, Pennsylvania, and West Virginia," July 30, 2010, http://www.netl.doe. gov/technologies/oil-gas/publications/ENVreports/ feoooo797-qpr-apr-jun2010.pdf.

10. U.S. Environmental Protection Agency website, "Water Trivia Facts," http://water.epa.gov/learn/kids/drinkingwater/water_trivia_facts.cfm.

11. Mielke, Diaz Anadon, and Narayanamurti, "Water Consumption of Energy Resource Extraction, Processing and Conversion."

12. Brian D. Lutz, Aurana N. Lewis, and Martin W. Doyle, "Generation, transport, and disposal of wastewater associated with Marcellus Shale gas development," *Water Resources Research* 49, 2 (February 2013), http://onlinelibrary.wiley.com/doi/10.1002/wrcr.20096/abstract.

13. Ralph Williams, "Solar cell makers, consider another (potentially) renewable resource: Water," RenewableEnergyWorld.com, June 17, 2011, http://www.renewableenergyworld.com/rea/news/article/2011/06/solar-cell-makers-consider-another-potentially-renewable-resource-water.

14. Don Hopey, "Gas drillers recycling more water, using fewer chemicals," *Pittsburgh Post-Gazette*, March 1, 2011, http://www.post-gazette.com/stories/local/region/gas-drillers-recycling-more-water-using-fewer-chemicals-210363/?p=2.

15. Ibid.

16. http://www.gasfrac.com/proven-proprietary-process.html.

Chapter 6: Criticisms of Fracking: Secret Chemicals

1. http://www.epa.gov/agriculture/lcra.html.
2. http://www.halliburton.com/public/projects/pubsdata/ Hydraulic_Fracturing/fluids_disclosure.html.
3. http://www.rangeresources.com/getdoc/50e3bc03-3bf6- 4517-a29b-e2b8ef0afe4f/Well-Completion-Reports.aspx.
4. http://boxofficemojo.com/movies/?id=gasland.htm.
5. http://www.rangeresources.com/rangeresources/ files/80/805df6fb-4b7b-4eb2-b39b-07fd1b2bb9ae.pdf.
6. http://www.rangeresources.com/rangeresources/ files/80/805df6fb-4b7b-4eb2-b39b-07fd1b2bb9ae.pdf.

Chapter 7: Criticisms of Fracking: Seismic Activity

1. http://www.ucl.ac.uk/archaeology/calendar/ articles/20100924.
2. http://www.mining-technology.com/projects/tautona_ goldmine/.
3. http://www.icdp-online.org/front_content.php?idcat=695.
4. http://www.icdp-online.org/front_content.php?idcat=695.
5. http://www.sakhalin-1.com/Sakhalin/Russia-English/ Upstream/media_news_events_Z44.aspx.
6. http://earthquake.usgs.gov/earthquakes/eqarchives/year/ eqstats.php.
7. http://www.osti.gov/geothermal/servlets/purl/895237- Vp8ett/895237.pdf.

8. http://esd.lbl.gov/research/projects/induced_seismicity/egs/geysers_why.html.

9. http://esd.lbl.gov/research/projects/induced_seismicity/egs/geysers.html.

10. http://earthquake.usgs.gov/earthquakes/world/events/1967_12_10.php.

11. http://www.telegraph.co.uk/news/worldnews/asia/china/4434400/Chinese-earthquake-may-have-been-man-made-say-scientists.html.

12. http://www.dailymail.co.uk/news/article-2412048/Fracking-DID-cause-109-earthquakes-Ohio-confirm-scientists-opposition-controversial-process-grows.html.

13. http://www.nbcnews.com/science/fracking-practices-blame-ohio-earthquakes-8C11073601.

14. http://www.sciencemag.org/content/161/3848/1301.short.

Chapter 8: Who Are the Anti-Fracking Activists?

1. http://books.google.ca/books?id=wwbwUDpoPrkC&pg=PA59&lpg=PA59&dq=the+last+year+has+seen+new+-discoveries+of+natural+gas+that+could+help+wean+us+off+dirtier+coal."&source=bl&ots=uO7eoh8dbO&sig=tYESgCbVD-AXPrIk1yFRnW3ZEn8&hl=en&sa=X&-ei=vhVdUpHJKsjAyAH_g4DoBw&ved=0CCwQ6AEwAA#v=onepage&q=the%20last%20year%20has%20seen%20new%20discoveries%20of%20natural%20gas%20that%20could%20help%20wean%20us%20off%20

dirtier%20coal."&f=false (*Eaarth: Making a Life on a Tough New Planet*, page 59).

2. Ibid., page 54.

3. http://content.time.com/time/health/ article/0,8599,1882700,00.html.

4. http://www.aoc.gov/press-room/capitol-power-plant- receives-permits-district-columbia-begin-cogeneration- project.

5. http://e360.yale.edu/feature/why_ill_get_arrested_to_ stop_the_burning_of_coal/2124/.

6. http://web.archive.org/web/20090316092947/http://www. capitolclimateaction.org/.

Chapter 9: Coal Mining vs. Fracking

1. "Gore defends his carbon credentials," Associated Press, February 28, 2007, http://www.nbcnews.com/id/17382210/ ns/us_news-environment/t/gore-defends-his-carbon- credentials/#.UedEkGQwbYc; "How much electricity does an American home use?" U.S. Energy Information Administration website, Frequently Asked Questions, accessed July 17, 2013, http://www.eia.gov/tools/faqs/faq. cfm?id=97&t=3.

2. 43rd Judicial District Court, Parker County, Texas: *Steven and Shyla Lipsky v. Durant et al.*, Order Denying Plaintiff's Sec. 27 Anti-Slapp Motion to Dismiss Range's Counter Claims, February 16, 2012, http://www. barnettshalenews.com/documents/2012/legal/Court%20

Order%20Denial%20of%20Lipsky%20Motion%20to%20
Dismiss%20Range%20Counterclaim%20-2-16-2012.pdf.

3. China Labor Bulletin: Deconstructing deadly details
from China's coal mine safety statistics. January 6, 2006.

4. Michael Martin, "Former miner explains culture of
mining," NPR, April 7, 2010, accessed July 17, 2013, http://
www.npr.org/templates/story/story.php?storyId=125676950.

5. United States Department of Labor, Mine Safety and
Health Administration, "Injury Trends in Mining,"
accessed July 17, 2013, http://www.usmra.com/saxsewell/
historical.htm.

Chapter 10: What Is Fracking Really Like in America?

1. http://blogs.wsj.com/wealth/2012/02/01/will-facebook-
really-create-1000-millionaires/.

2. http://www.bls.gov/lau/.

3. http://www.bls.gov/eag/eag.ma.htm.

4. http://www.stats.gov.sk.ca.

5. http://www.manhattan-institute.org/html/gpr_01.htm#.
UloN2hbtX4Y.

6. http://www.gov.ns.ca/finance/statistics/analysis/default.
asp?id=23.

7. Michael Grunwald, "Why our farm policy is failing,"
Time, November 2, 2007.

8. Ibid.

9. "Natural-gas royalties could top $1.2 billion in Pa.,"
Associated Press, January 29, 2013, accessed July 29, 2013,

http://articles.philly.com/2013-01-29/business/36637967_1_royalty-payments-royalty-owners-natural-gas.

10. Jake Whitman and Sharyn Alfonsi, "Oil boom fueling fortunes in Kansas," ABC News, May 10, 2012, http://abcnews.go.com/Business/MadeInAmerica/oil-boom-creating-overnight-millionaires-kansas-us-energy/story?id=16297143, accessed July 29, 2013.

11. Phil Davies, "Sand surge," FedGazette, The Federal Reserve Bank of Minneapolis, July 16, 2012, accessed July 29, 2013, http://www.minneapolisfed.org/publications_papers/pub_display.cfm?id=4921.

12. David Bailey, "In North Dakota, hard to tell an oil millionaire from regular Joe," Reuters, October 3, 2012, accessed July 29, 2013, http://www.reuters.com/article/2012/10/03/us-usa-northdakota-millionaires-idUSBRE8921AF20121003.

13. Malia Spencer, "Marcellus royalties rising," *Pittsburgh Business Times*, June 10, 2013, accessed July 29, 2013, http://www.bizjournals.com/pittsburgh/blog/energy/2013/06/marcellus-royalties-rising.html.

14. http://www.project-syndicate.org/blog/frack-to-the-future.

15. http://www.eia.gov/todayinenergy/detail.cfm?id=7350#tabs_co2emissions-1.

16. http://www.census.gov/population/estimates/nation/intfile3-1.txt.

Chapter 11: Shale Gas Around the World

1. http://www.eia.gov/analysis/studies/worldshalegas/.
2. http://www.bls.gov/web/laus/laumstrk.htm.
3. William Manchester and Paul Reid, *The Last Lion: Winston Spencer Churchill: Defender of the Realm*, 1940–1965 (New York: Little, Brown and Co., 2012).
4. Michael Sontheimer, "Germany's WWII Occupation of Poland: 'When We Finish, Nobody Is Left Alive,'" Der Spiegel Online, International, May 27, 2011, http://www.spiegel.de/international/europe/germany-s-wwii-occupation-of-poland-when-we-finish-nobody-is-left-alive-a-759095.html.
5. Andrew Purvis, "Remaking Poland," *Time*, April 9, 2008.
6. *Shale gas report – Poland*, Ernst & Young, 2012, http://www.ey.com/Publication/vwLUAssets/Shale_gas_report_-_Poland/$FILE/Shale_gas_report%E2%80%94Poland.pdf.
7. "Mad and messy regulation," *The Economist*, July 10, 2013, http://www.economist.com/blogs/easternapproaches/2013/07/shale-gas-poland.
8. Gazprom Export, Country Profile: Poland, http://www.gazpromexport.ru/en/partners/poland/.
9. http://www.eia.gov/dnav/ng/hist/n9190us3A.htm
10. Natural gas in the U.S. and Canada is typically measured in British thermal units (Btu) or thousand cubic feet (mcf). One mcf of gas is approximately equal to a million Btus, or one MMBtu. There are approximately 35 cubic feet in one cubic metre.

11. http://www.sabinepipeline.com/Home/Report/tabid/241/default.aspx?ID=52.

12. http://www.naturalgaseurope.com/russia-and-poland-agrees-on-gas-price-reduction.

13. Sergei Ispolatov, *Izvestia*, February 1, 2013, http://izvestia.ru/news/544100.

14. http://en.rian.ru/russia/20071101/86223448.html.

15. James Marson and Marynia Kruk, "Gazprom's pipeline move meets with disbelief," *The Wall Street Journal*, April 5, 2013, http://online.wsj.com/article/SB100014241278873239163045784044509798448o8.html.

16. "Mad and messy regulation," *The Economist*, July 10, 2013, http://www.economist.com/blogs/easternapproaches/2013/07/shale-gas-poland.

17. Ladka Bauerova, "Chevron draws Europe toward natural gas independence: Energy," Bloomberg, July 24, 2013, http://www.bloomberg.com/news/2013-07-23/chevron-anticipates-europeans-prefer-fracking-to-putin-energy.html.

18. "Mad and messy regulation," *The Economist*.

19. http://www.eia.gov/countries/country-data.cfm?fips=UP.

20. http://www.world-nuclear.org/info/Country-Profiles/Countries-T-Z/Ukraine/#.Udh5pj7F3LN.

21. http://www.economist.com/node/12903050.

22. http://www.telegraph.co.uk/news/worldnews/1562838/Yushchenko-Russia-blocking-poisoning-probe.html.

23. http://www.dailymail.co.uk/news/article-1051871/Stalins-mass-murders-entirely-rational-says-new-Russian-textbook-praising-tyrant.html; http://www.reuters.com/article/2013/03/05/us-russia-stalin-idUSBRE92400120130305.

24. http://www.bbc.co.uk/news/world-asia-22737548; http://www.cbc.ca/news/world/story/2011/05/30/germany-nuclear-shutdown.html.

25. http://www.energydelta.org/mainmenu/energy-knowledge/country-gas-profiles/country-profile-germany.

26. http://www.telegraph.co.uk/finance/newsbysector/energy/10079798/IEA-warns-Germany-on-soaring-green-dream-costs.html.

27. http://www.bloomberg.com/news/2013-02-08/german-shale-gas-fracking-rules-sought-in-merkel-coalition-paper.html.

28. http://www.independent.co.uk/news/world/europe/nicolas-sarkozy-did-take-50-million-of-muammar-gaddafis-cash-french-judge-is-told-8435872.html.

29. http://www.telegraph.co.uk/news/worldnews/africaandindianocean/algeria/9831931/Algeria-gas-pipeline-attack-kills-two-guards.html.

30. http://news.yahoo.com/2-guards-killed-attack-algerian-gas-pipeline-130359304-finance.html.

31. World Shale Gas Resources, April 2011, U.S. Energy Information Administration.

32. http://www.vermilionenergy.com/operations/france/english/shale-oil.cfm.

33. http://www.time.com/time/magazine/article/0,9171,901020325-218398,00.html.

34. http://www.businessweek.com/magazine/content/11_15/b4223060759263.htm.

35. http://www.france24.com/en/20130226-french-unemployment-level-hits-15-year-high.

36. http://www.businessweek.com/news/2011-10-04/france-to-keep-fracking-ban-to-protect-environment-sarkozy-says.html.

37. https://www.wind-watch.org/allies.php#uk.

38. http://www.dailymail.co.uk/news/article-2013233/The-wind-turbine-backlash-Growing-public-opposition-thwarts-green-energy-drive.html.

39. http://www.thetimes.co.uk/tto/business/industries/naturalresources/article3683377.ece.

40. http://royalsociety.org/uploadedFiles/Royal_Society_Content/policy/projects/shale-gas/2012-06-28-Shale-gas.pdf.

41. http://www.independent.co.uk/environment/green-living/exclusive-fracking-company-we-caused-50-tremors-in-blackpool-but-were-not-going-to-stop-6256397.html.

42. http://www.dailymail.co.uk/news/article-1393033/Man-earthquake-strikes-Blackpool-consequences-severe-UKs-gas-drilling-industry.html.

43. http://www.thesun.co.uk/sol/homepage/news/3504143/Blackpool-rocked.html.

44. "Putin slams NATO on Libya attacks," RIA Novosti, April 26, 2011, http://en.rian.ru/world/20110426/163721016.html.

45. U.S. Energy Information Administration, "Technically Recoverable Shale Oil and Shale Gas Resources: An Assessment of 137 Shale Formations in 41 Countries Outside the United States," June 2013, http://www.eia.gov/analysis/studies/worldshalegas/pdf/fullreport.pdf?zscb=45256562.

46. Energy Delta Institute, Country Profiles, Bulgaria, http://www.energydelta.org/mainmenu/energy-knowledge/country-gas-profiles/bulgaria.

47. "Bulgaria cancels Chevron shale gas permit," Reuters, Jan. 17, 2012, http://www.reuters.com/article/2012/01/17/us-bulgaria-shalegas-chevron-idUSTRE80G18J20120117.

48. John Daly, "Russia Behind Anti-Fracking Protests?" Oilprice.com, February 4, 2012, http://oilprice.com/Energy/Natural-Gas/Russia-Behind-Bulgarian-Anti-Fracking-Protests.html.

49. "Thousands protest Bulgaria 'fracking' plans," Agence France-Presse, January 14, 2012, http://www.rawstory.com/rs/2012/01/14/thousands-protest-bulgaria-fracking-plans/.

50. John Daly, "Russia Behind Anti-Fracking Protests?" Oilprice.com, February 4, 2012, http://oilprice.com/Energy/Natural-Gas/Russia-Behind-Bulgarian-Anti-Fracking-Protests.html.

51. "Bulgaria wants safe technology to explore shale gas reserves," Voice of Russia, TASS, February 5, 2013, http://voiceofrussia.com/2013_02_05/Bulgaria-wants-safe-technology-to-explore-shale-gas-reserves/.

52. Paul Waldie, "How fracking weakens Gazprom, the bedrock beneath Putin's feet," *The Globe and Mail*, Feb. 18, 2013, http://www.theglobeandmail.com/report-on-business/industry-news/energy-and-resources/how-fracking-weakens-gazprom-the-bedrock-beneath-putins-feet/article8791722/.

53. Global Museum of Communism, Lithuania Exhibit, History, http://lithuania.globalmuseumoncommunism.org/lithuania/history.

54. James Kanter, "At anchor off Lithuania, its own energy supply," *The New York Times*, July 4, 2013, http://www.

nytimes.com/2013/07/05/business/energy-environment/
lithuania-aims-for-energy-independence.
html?pagewanted=all.

55. Bryan Bradley, "Lithuanian gas utility Dujos renegotiat-
ing Gazprom supply deal," Bloomberg, July 31, 2013,
http://www.bloomberg.com/news/2013-07-31/lithuanian-
gas-utility-dujos-renegotiating-gazprom-supply-deal.html.

56. Juris Kaža, "Lithuania, Gazprom see cheaper gas price
after meeting," The Wall Street Journal, September 6,
2013, http://online.wsj.com/article/BT-CO-20130906-
707942.html.

57. Gazprom press release, September 6, 2013, http://www.
gazprom.com/press/news/2013/september/article170668/.

58. Neil Buckley, "Romania and Lithuania back fracking,"
Financial Times, February 5, 2013, http://www.ft.com/intl/
cms/s/0/fa2812bc-6fa6-11e2-956b-00144feab49a.
html#axzz2eLWQ7yEm.

59. Bryan Bradley, "Chevron gets Lithuanian president's
backing for shale gas," Bloomberg, February 4, 2013,
http://www.bloomberg.com/news/2013-02-04/chevron-
gets-lithuanian-president-s-backing-for-shale-gas.html.

60. James Kanter, "At anchor off Lithuania, its own energy
supply," The New York Times, July 4, 2013, http://www.
nytimes.com/2013/07/05/business/energy-environment/
lithuania-aims-for-energy-independence.
html?pagewanted=all.

61. Danielle Kurtzleben, "Report: America lost 2.7 million
jobs to China in 10 years," U.S. News & World Report,
August 24, 2012, http://www.usnews.com/news/

articles/2012/08/24/report-america-lost-27-million-jobs-to-china-in-10-years.

62. James R. Hagerty, "U.S. manufacturers gain ground," *The Wall Street Journal*, August 18, 2013, http://online.wsj.com/article/SB10001424127887323423804579020732661092434.html.

63. Ibid.

64. Christopher Alessi and Stephanie Hanson, "Expanding China-Africa Oil Ties," Council on Foreign Relations, February 8, 2012, http://www.cfr.org/china/expanding-china-africa-oil-ties/p9557.

65. Colin McClelland and Bradley Olsen, "Sinopec buys Canada's Daylight for $2.1 billion to gain shale-gas assets," Bloomberg, October 10, 2011, http://www.bloomberg.com/news/2011-10-09/sinopec-agrees-to-buy-daylight-energy-for-2-1-billion-to-meet-fuel-demand.html.

66. Shawn McCarthy and Carrie Tait, "With $15.3-billion takeover complete, CNOOC to rely on Nexen managers," *The Globe and Mail*, February 27, 2013, http://www.theglobeandmail.com/report-on-business/industry-news/energy-and-resources/with-153-billion-takeover-complete-cnooc-to-rely-on-nexen-managers/article9134322/.

67. "Sinopec, Conoco Shale Gas Venture," Zacks Equity Research, December 31, 2012, http://finance.yahoo.com/news/sinopec-conoco-shale-gas-venture-221856811.html.

68. "Sinopec, Conoco Shale Gas Venture," Zacks Equity Research; Simon Hall, "Total extends its China ties," *The Wall Street Journal*, March 18, 2012, http://online.wsj.com/

article/SB10001424052702304636404577288961610767188.
html.

69. Stefan Nicola and Rainer Buergin, "German shale-gas
    fracking rules sought in Merkel coalition paper,"
    Bloomberg, February 8, 2013, http://www.bloomberg.
    com/news/2013-02-08/german-shale-gas-fracking-rules-
    sought-in-merkel-coalition-paper.html.

70. http://www.bbc.co.uk/news/world-middle-east-17808954.

71. http://www.oilinisrael.net/oil-in-israel-articles/haifa-gas-
    discovery-bumped-to-5-trillion-cubic-feet.

72. http://blogs.marketwatch.com/energy-ticker/2013/04/04/
    noble-raises-reserve-estimates-on-israels-tamar-gas-field/.

73. http://www.bloomberg.com/news/2013-03-30/israel-
    begins-gas-production-at-tamar-field-in-boost-to-
    economy.html.

74. http://www.bloomberg.com/news/2013-03-30/israel-
    begins-gas-production-at-tamar-field-in-boost-to-
    economy.html.

75. http://maya.tase.co.il/bursa/report.asp?report_
    cd=745484&CompCd=600.

76. http://www.algemeiner.com/2013/04/12/tamar-gas-field-
    operator-looking-to-build-pipeline-from-israel-to-turkey/.

77. http://jordantimes.com/arab-potash-company-looks-to-
    import-natural-gas-from-israel.

78. http://www.globes.co.il/serveen/globes/docview.
    asp?did=1000832632&fid=1725.

79. Stikeman Elliot, Canadian Energy Law Blog, "Quebec
    and federal government enter St. Lawrence offshore oil
    deal," April 6, 2011, http://www.canadianenergylaw.

com/2011/04/articles/oil-and-gas/quebec-and-federal-government-enter-st-lawrence-offshore-oil-deal/.

80. Leger Marketing, "Study on the Knowledge and Perceptions of Quebecers About Energy," August 2012.

81. Elizabeth Thompson, "Quebec soft on crime, senator charges as province, feds spar over crime bill," iPolitics, November 16, 2011, http://www.ipolitics.ca/2011/11/16/quebec-soft-on-crime-conservative-charges-as-province-feds-spar-over-crime-bill/.

82. Richard J. Brennan, "Majority of Canadians support return of death penalty, poll finds," *Toronto Star*, February 8, 2012, http://www.thestar.com/news/canada/2012/02/08/majority_of_canadians_support_return_of_death_penalty_poll_finds.html.

83. *Canadian Parliamentary Review*, March 22, 2012, http://www.thefreelibrary.com/Quebec.-a0285885588.

84. Rhéal Séguin and Bertrande Marotte, "Quebec opposition denounces fracking legislation," *The Globe and Mail*, May 15, 2013, http://www.theglobeandmail.com/news/national/quebec-opposition-denounces-fracking-legislation/article11943093/.

85. Sophie Cousineau, Bertrande Marotte, and Rhéal Séguin, "Quebec gas in peril as PQ signals ban," *The Globe and Mail*, September 20, 2012, http://www.theglobeandmail.com/report-on-business/industry-news/energy-and-resources/quebec-gas-in-peril-as-pq-signals-ban/article4557380/.

86. Nicolas Van Praet, "Former Quebec premier says province botched shale gas start," *National Post*, October 23,

2012, http://business.financialpost.com/2012/10/23/former-quebec-premier-says-province-botched-shale-gas-start/?__lsa=28b0-6009.

87. Jeff Gray, "U.S. firm to launch NAFTA challenge to Quebec fracking ban," *The Globe and Mail*, November 15, 2012, http://www.theglobeandmail.com/globe-investor/us-firm-to-launch-nafta-challenge-to-quebec-fracking-ban/article5337929/.

88. Nicolas Van Praet, "Fighting Quebec's anti-energy myths," *National Post*, October 23, 2012, http://business.financialpost.com/2012/10/23/fighting-quebecs-anti-energy-myths/?__lsa=28b0-6009.

89. "Le Cégep de Thetford Offre une Formation de Classe Mondiale Pour les Québécois," Cégep de Thetford, news release, May 23, 2012, http://www.cegepth.qc.ca/public/e78f0aa9-d295-4d45-a478-2b13cc3478f3/communiques/2011-2012/communique_de_presse_finale_%282%29.pdf.

90. Patricia Laya, Esteban Duarte and Manuel Baigorri, "Repsol LNG sale seen delayed as shale goring Canaport," Bloomberg, February 15, 2013, http://www.bloomberg.com/news/2013-02-15/repsol-lng-sale-seen-delayed-as-shale-goring-canaport.html.

91. Repsol website, "Canaport Facts," http://www.repsolenergy.com/canaport/canaport_facts.html.

92. Jeff Lewis, "Canaport undermined by shale gas prices; $1.3B writedown," *National Post*, February 27, 2013, FP3.

93. Province of New Brunswick, "The New Brunswick Oil and Natural Gas Blueprint," May 2013, http://www2.gnb.

ca/content/dam/gnb/Departments/en/pdf/Publications/
9281%20ONG%20English%20Final%20web.pdf.

94. "Shale gas truck seized by 'native warriors' in N.B.," CBC
News, June 4, 2013, http://www.cbc.ca/news/canada/new-
brunswick/story/2013/06/04/nb-swn-elsipogtog-truck-
rcmp.html; "RCMP probe work crew property damage,
trees blocking road," CBC News, June 24, 2013,
http://www.cbc.ca/news/canada/new-brunswick/
story/2013/06/24/nb-trees-cut-block-crew.html; Claire
Stewart-Kanigan, "Midnight confiscation of drilling
equipment at New Brunswick anti-fracking protest,"
rabble.ca, June 25, 2013, http://rabble.ca/news/2013/06/
midnight-confiscation-drilling-equipment-new-
brunswick-anti-fracking-protest; Angela Giles, "Peaceful
resistance to shale gas in N.B. may soon come to a head,"
rabble.ca, June 14, 2013, http://rabble.ca/blogs/bloggers/
council-canadians/2013/06/peaceful-resistance-to-shale-
gas-nb-may-soon-come-to-head ; Brent Patterson, "RCMP
make arrests at New Brunswick fracking protest this
morning," rabble.ca, June 14, 2013, http://rabble.ca/blogs/
bloggers/brent-patterson/2013/06/rcmp-make-arrests-new-
brunswick-fracking-protest-this-morning.

95. David P. Ball, "Fracking troubles Atlantic First Nations
after two dozen protestors arrested," Indian Country
Today Media Network, June 28, 2013, http://
indiancountrytodaymedianetwork.com/2013/06/28/
fracking-troubles-atlantic-first-nations-after-two-dozen-
protesters-arrested-150192.

Chapter 12: *Gasland* and Josh Fox

1. http://dailycaller.com/2010/04/22/james-cameron-is-the-ultimate-environmentalist-except-when-he-flies-on-pollution-spewing-private-jets/.
2. http://www.boxofficemojo.com/movies/?id=promisedland2012.htm.
3. http://www.eia.gov/countries/country-data.cfm?fips=TC.
4. http://www.youtube.com/watch?v=Aa1FQHywwp8.
5. http://www.youtube.com/watch?v=-8R4sbchSao&feature=youtube.
6. http://www.blueflowerarts.com/josh-fox.
7. http://artsbeat.blogs.nytimes.com/2010/06/21/a-muckraker-targets-onshore-drilling/.
8. http://www.internationalwow.com/newsite/josh.html.
9. http://www.cinevegas.com/blog/?p=38.
10. http://www.thawaction.org.
11. http://www.timeout.com/newyork/theater/youre-in-the-army-now.
12. http://www.nytimes.com/gwire/2011/02/24/24greenwire-groundtruthing-academy-award-nominee-gasland-33228.html?sq=gasland&st=cse&scp=1&pagewanted=all.
13. http://www.youtube.com/watch?v=M9gMOJpyYzo&feature=youtube.
14. https://www.youtube.com/watch?feature=player_embedded&v=M9gMOJpyYzo; http://www.nwpoa.info.
15. http://blogs.ft.com/energy-source/2011/02/18/gazprom-chief-steps-up-attacks-on-shale-gas/#axzz2LgEbbq3n.

16. https://twitter.com/VzlaEmbassyUS/
    status/42294129748754432.

17. http://venezuela-us.org/2011/02/24/venezuelan-participates-
    in-documentary-nominated-to-academy-awards/.

18. http://www.eluniversal.com/economia/130710/venezuelas-
    oil-exports-to-the-us-fall-in-2013.

19. http://water.epa.gov/type/groundwater/uic/class2/
    hydraulicfracturing/wells_hydroreg.cfm.

20. http://www.nytimes.com/gwire/2011/02/24/24greenwire-
    groundtruthing-academy-award-nominee-gasland-33228.
    html?sq=gasland&st=cse&scp=1&pagewanted=all.

21. http://democrats.energycommerce.house.gov/sites/
    default/files/documents/Hydraulic%20Fracturing%20
    Report%204.18.11.pdf.

22. http://www.nytimes.com/gwire/2011/02/24/24greenwire-
    groundtruthing-academy-award-nominee-gasland-33228.
    html?sq=gasland&st=cse&scp=1&pagewanted=all.

23. http://www.huffingtonpost.com/2011/08/22/halliburton-
    executive-drinks-fracking-fluid_n_933621.html.

24. http://www.epa.gov/ttnatw01/hlthef/ethylben.html.

25. http://www.epa.gov/oppsrrd1/REDs/tcmtb_red.pdf.

26. http://www.propublica.org/article/hydrofracked-one-
    mans-mystery-leads-to-a-backlash-against-natural-gas-
    drill/single.

27. http://www.counterpunch.org/2011/03/25/fracking-the-
    wind-river-country/.

28. http://cogcc.state.co.us/library/GASLAND%20DOC.pdf.

29. http://www.cancer.gov/cancertopics/wyntk/pancreas/
    page4.

30. http://cogcc.state.co.us/Library/Presentations/Glenwood_ Spgs_HearingJuly_2009/(2_B)_ InvestigationintoComplaintofNewGasSeep.pdf.

31. http://cogcc.state.co.us/cogis/ComplaintReport.asp?doc_ num=200191771.

32. http://articles.philly.com/2012-08-27/news/33403570_1_ susquehanna-county-town-cabot-oil-baby-drill.

33. http://dimockproud.files.wordpress.com/2013/02/exhibit-d-to-consent-order-12-14-2010.pdf.

34. http://uk.reuters.com/article/2012/05/11/usa-fracking-dimock-idUKL1E8GBVGN20120511.

35. http://www.youtube.com/watch?v=1QGAKe8ttnw.

36. http://dimockproud.files.wordpress.com/2013/02/exhibit-d-to-consent-order-12-14-2010.pdf.

37. http://www.nrdc.org/about/finances.asp.

38. http://www.hangerforgovernor.com/about_john_hanger.

39. http://dimockproud.com.

40. http://nyhistoric.com/2012/10/burning-springs/.

41. http://www.niagarafrontier.com/burningsprings.html.

42. http://query.nytimes.com/mem/archive-free/pdf?res= F30716F63E5C15738DDDA80A94D9405B8484F0D3.

43. Carl Zimmer, "Science and Politics: The Tale of George Washington's Swamp Gas," *Discover*, October 17, 2008, http://blogs.discovermagazine.com/loom/2008/10/17/ science-and-politics-the-tale-of-george-washingtons-swamp-gas/#.UhrS7mQwbYc.

44. Benjamin Franklin and Jared Sparks, *The Works of Benjamin Franklin* (Hillard, Gray, 1840).

45. http://cogcc.state.co.us/library/GASLAND%20DOC.pdf.

46. Jack Z. Smith, "Owner of contaminated water well in
    Parker County loses in court," *Arlington Star-Telegram*,
    February 18, 2012, http://www.star-telegram.
    com/2012/02/17/3744111/owner-of-contaminated-water-
    well.html.

Conclusion

1. http://www.forbes.com/sites/christopherhelman/2013/07/27/
   father-of-the-fracking-boom-dies-george-mitchell-urged-
   greater-regulation-of-drilling/.
2. http://thehill.com/blogs/e2-wire/e2-wire/319649-
   booming-oil-production-boosted-gdp-estimate-white-
   house-advisers-say.
3. http://www.aei-ideas.org/2013/11/some-mind-blowing-
   facts-about-us-metro-economies/.
4. http://www.reuters.com/article/2013/10/07/petronas-
   canada-idUSL4N0HX05820131007.
5. http://rt.com/business/poland-shale-gas-fracking-
   europe-154/.
6. http://www.telegraph.co.uk/finance/comment/
   ambroseevans_pritchard/10257988/Polands-shale-drive-
   will-transform-Europe-if-it-does-not-drop-the-ball.html.
7. http://rt.com/op-edge/fracking-technologies-climate-
   change-085/.